To Dad —
with love from Toni
1/31/76

Minerals: Nature's Fabulous Jewels

Arthur Court

Ian Campbell
President, California Academy of Sciences
Professor Emeritus of Geology, California Institute of Technology

PHOTOGRAPHS BY **M. Halberstadt**

NERALS
NATURE'S FABULOUS JEWELS

Harry N. Abrams, Inc., *Publishers*, New York

Frontispiece: QUARTZ (variety agate)

Collection: Court
Size: 2×4 in. (5×10 cm.)
Locality: South Africa
SiO_2 Hexagonal

Quartz in fine-grained, massive form is known as chalcedony, and chalcedony, when banded, is called agate. The banding results from slow deposition of silica, in parallel or concentric layers, each of which may incorporate small, slightly different amounts of some colorful impurity or may vary only slightly in submicroscopic porosity. In either case, color patterns may be enhanced if we soak the agates for long periods in an appropriate chemical. One ancient method, still in use, is to soak them in honey (or a sugar solution), followed by a soaking in sulfuric acid, which converts the sugar to carbon and produces various shades of grays, browns, and blacks within the bands.

Bitita Vinklers   *Editor*
Margaret L. Kaplan   *Managing Editor*
Quentin Fiore   *Book Design*
Nai Y. Chang   *Vice-President, Design*

Library of Congress Cataloging in Publication Data
Campbell, Ian, 1899–
    Minerals: nature's fabulous jewels.

    Bibliography: p.
    1. Mineralogy. I. Court, Arthur, joint author.
II. Title.
QE363.2.C28   549   74-6267
ISBN 0-8109-0311-3

CONTENTS

Acknowledgments

That a book such as this must involve cooperation and assistance from many must be immediately obvious, and it is clearly impossible to name all to whom we are indebted. However, we gratefully extend special appreciation to the following discriminating collectors who generously loaned, for photographing, some of their most prized specimens:

Ruth Brown, Reno, Nevada
The California Academy of Sciences, San Francisco, California
The California Division of Mines and Geology, San Francisco, California
The late Mr. Brooks Davis, and Mrs. Davis, Tucson, Arizona
M. Halberstadt, San Francisco, California
Jack Halpern, San Francisco, California
Mr. and Mrs. Clifford Krueger, San Francisco, California
Lewis K. Land, San Francisco, California
A. L. McGuinness, San Mateo, California
Steven Smale, Kensington, California

In addition, it should be pointed out that many of the featured specimens are from Arthur Court's own extensive collection.

Ian Campbell gratefully acknowledges his indebtedness to the late Charles Palache, since much of whatever virtue may lie in the text pages results from his inspired tutelage of many years ago. And we both want to express our appreciation for the thorough and helpful editing by Bitita Vinklers, which has done much to improve the accuracy and the readability of this book.

Arthur Court
Ian Campbell

INTRODUCTION

Man delights in his natural world. Throughout history he has been exploring it, sampling it, codify- ing it, and seeking a better understanding of it. Long ago, he simply but effectively classified the natural world into three great kingdoms: the animal kingdom, the plant kingdom, and the mineral kingdom.

The animal kingdom comprises many thousands—in fact millions—of species, most of them now extinct and known solely from the fossil record. Of the many living species of animals, man has domesticated only a few—such as the horse, the cow, the sheep, the dog—but many of these species he has greatly improved through selective breeding procedures.

The plant kingdom comprises many more thousands of species, living and vanished. Again, man has domesticated only a few, but some of these too he has spectacularly improved through the years —such as corn, wheat, tomatoes.

The mineral kingdom, in contrast, comprises only some 2500 or fewer "species," almost none of which, so far as now known, has vanished from the earth. Of these 2500 species, man has "domes-ticated" (that is, turned to his own use) a far larger proportion—numbering about 200 species—than he has from the animal and plant kingdoms. Another interesting point is that, while man has not "bred" minerals, he has synthesized many, including some gemstones. Yet, in contrast to his success in breeding improved varieties of animals and vegetables, man has barely succeeded, in these syn-theses, in equaling nature's product—very seldom has he improved upon it, and then only slightly.

If bread was "the staff of life" of our ancestors, today the staff of life is steel (or perhaps alumi-num), for without the agricultural machinery produced from minerals, man could not raise enough bread to put into the mouths of the more than three billion people in the world, nor could he feed or milk the cows required to provide milk for the rapidly increasing world population.

Today there are many people who may never have seen a cow or a wheat field, much less a mine—and who are unfamiliar with minerals. For all those who as yet know almost nothing about minerals, as well as for many of those who already know a little, the aim of this book is to present in very simple terms something of the beauty, the utility, and the scientific interest that lie within the mineral kingdom—the kingdom on which so much of our lives today depends.

Man's Need for Minerals

It has been said that "minerals are man's best friends." If a good life, in the best figurative sense, depends on friends, then in a literal sense the statement is true—minerals *are* man's best friends. Without minerals man in all probability would still be living in caves.

To be sure, food (the product of the animal and vegetable kingdoms) may seem more essential to life than the products of the mineral kingdom—but for life as we live it today minerals are tremendously important. The daily requirements of food calories of twentieth-century man are very likely no more (they may well be less) than were the requirements of our caveman ancestors. In contrast, man's per-capita consumption of minerals has been constantly increasing since he first discovered that certain kinds of stones made superior arrowheads and that certain other kinds made superior grinding mortars. Such advances and man's changing demands for minerals are evidenced by the categories of the Stone Age, the Bronze Age, the Iron Age, the Age of Steel and Concrete for today's world, and the Nuclear Age for the future. All of the materials referred to are either mined or are mineral products.

We are rightly concerned that much of the world's population today subsists on a diet of less than 1500 calories per day. Yet the difference between this figure and the 2500 to 3000 calories considered adequate for most adults is not so appalling as is the fact that there is a difference of twenty to fifty times in man's "mineral diet" between the developed and the less developed countries of the world. For example, the average yearly per-capita consumption of copper in Europe and North America in 1969 was slightly over 14 pounds; in the less developed countries, it was just over $\frac{1}{4}$ pound. In the case of iron, the consumption was 838 pounds per capita in the developed countries, and only 24 in the less developed. And for such an important fertilizer component as is provided by potash minerals, the figures were 36 pounds versus $\frac{1}{2}$ pound. Can "one world" long exist with such disparity? And where will supplies be found to meet man's insatiable demands upon the mineral kingdom?

Compounding this problem is the fact that, along with the great diversity in consumption demands by countries, there is even greater diversity in the natural distribution of minerals; this creates one of the most intriguing and challenging aspects of the mineral world.

Every square mile of the earth's crust offers a somewhat different—sometimes a startlingly different—array of minerals (in kind or in amount) from every other square mile. For instance, in the United States, within the 3,022,387 square miles of the forty-eight contiguous states, there is virtually nowhere a concentration of tin-bearing minerals great enough to justify commercial development. Yet Americans are the greatest users of tin in the world! So, to supply their needs they must depend on such far-off places as Bolivia and Indonesia. In addition, the United States produces almost no long-fiber asbestos, a tremendously important industrial mineral used in everything from brake bands to fireproof clothing and vinyl tile. Yet, just across the border in our good neighbor, Canada, are some of the world's greatest deposits of asbestos. The aluminum industry of the entire world was for

2. QUARTZ (variety rock crystal)

Collection: Court
Size: $5\frac{1}{2} \times 12$ in. (14×30 cm.)
Locality: Mount Ida, Arkansas
SiO_2 Hexagonal

Although quartz has the seemingly simple composition of silicon dioxide, SiO_2, chemists and mineralogists are divided on whether it should be classed with the oxide group of minerals or—if it is viewed as a silicate of silicon—as a silicate. Structural crystallographic evidence gives weight to the latter view. In any event, the mineral is one of considerable scientific interest, economic importance, and aesthetic appeal.

many years dependent on a rare mineral, cryolite (a sodium aluminum fluoride), obtained from a deposit in southern Greenland containing millions of tons. Elsewhere in the world cryolite, though long and eagerly sought for, has been found in only about half a dozen localities, and in none of these has more than a few pounds of it been present.

If geologists and mineralogists face difficult problems in trying to explain the strikingly localized occurrences of important minerals, these are more than matched by the problems that economists, industrialists, and diplomats have in trying to achieve equitable production and distribution of minerals. Such considerations surely make mineralogy and mineral economics not only a fascinating but an obligatory study for the intelligent layman. Although in many respects the United States is far from being a have-not nation, it is dependent on foreign imports for its mineral needs to a far greater degree than many people realize.

Nevertheless, before we go any further into the broader problems of minerals—where they are and why—we should make sure that we know just what we are talking about. What *is* a mineral?

What Is a Mineral?

What is a mineral? The dictionary tells us that a mineral is a naturally occurring, homogeneous, inorganic substance with a definite chemical composition possessing a distinctive crystalline structure. To understand this properly, let us consider these properties one by one.

Naturally occurring. This characteristic immediately rules out bricks, pennies, and a whole host of man-made mineral products, including synthetic gems, which—except for the fact that they are man-made—are virtually identical with their namesakes from the natural world. In other words, a mineral must be a product of nature, created on or within the earth (though a few new minerals have been discovered on the moon), not in a factory or through some process designed by man.

Homogeneous. This is a word we are perhaps very familiar with, because we see it so often on milk cartons. If milk has been homogenized it will no longer, on being allowed to stand, separate into a layer of cream on top and a layer of thin milk below. A sample drawn at any time from any level within a carton or even a tank of homogenized milk will be identical to all other samples that might be taken from the carton or the tank; not only will each sample look the same, but on analysis it will yield the same percentage of butter fat, the same amounts of calcium and other mineral nutrients, and the same amount of water. So, if a naturally occurring substance, such as a mineral, is homogeneous, a sample from any portion of it is essentially identical—particularly in its chemical composition and in its crystalline structure—to a sample from any other portion.

Among other things, the requirement for homogeneity answers the often-heard question "Is a rock a mineral?" The answer is usually "No." We have all seen polished granite and noted the different colors displayed on the surface—such as pink, white, black. So granite is not homogeneous, and in fact each of the colors represents a different mineral, which in itself *is* homogeneous: pink feldspar, white quartz, black mica—each of these is a separate and distinct mineral. But we have all also seen pure white marble. This is homogeneous, and such marble is both a rock *and* a mineral—usually calcite.

Inorganic. Here we can once again, if we wish, apply the concept of animal, plant, and mineral categories. Organic substances include both animals and plants; inorganic material is here equated with the mineral world. An animal product, such as a pearl produced by an oyster, is therefore not truly a mineral. Coal, a fossilized plant product, is likewise not a mineral, although coal is often obtained from underground mines, and some very hard, lustrous black varieties yield the semiprecious stone known as jet.

A definite chemical composition. The quality of homogeneity in itself implies something of the chemistry of minerals. A homogeneous substance cannot be too diverse chemically. Each mineral has a very definite chemical composition; just one example is sodium chloride—known as halite to mineralogists, NaCl to chemists, and common salt to everyone. From a mass of halite, any portion will yield the same analysis to a chemist: a proportion of one atom of sodium (Na) to one atom of chlorine (Cl). It is therefore homogeneous, and it obviously has a definite chemical composition.

It is true that many minerals contain small amounts of impurities, usually present in only fractions of 1 percent of the bulk composition. These are sometimes responsible for unusual colors or internal textures and may thereby add to the aesthetic value, and to a dealer's prices! It is worth noting, too, that ratios of certain components within a mineral may vary without destroying the over-all homogeneity; in such cases, the range of chemical variation can occur only within well-defined limits. Hence the requirement that a mineral must possess a definite chemical composition remains valid. An example is afforded by a rather widely distributed, green mineral called olivine (gem-quality olivine is known as peridot, and was particularly prized by ancient Egyptians). It is composed of iron (Fe), magnesium (Mg), silicon (Si), and oxygen (O). The chemist writes his formula for olivine as $(Fe, Mg)_2SiO_4$, which is his shorthand to indicate that some olivine may be Fe_2SiO_4, some may be Mg_2SiO_4, and some may be a mixture in any proportion from 100 percent Fe_2SiO_4 to 100 percent Mg_2SiO_4. Most olivine does in fact contain both magnesium and iron, with magnesium usually predominating in a ratio of 5 or 6 to 1. The "pure" Mg_2SiO_4 and the "pure"

Fe_2SiO_4 are referred to as end members of the olivine series. They are all olivines, but the end members constitute the defined limits of this chemical system.

A distinctive crystalline structure. The crystalline structure is perhaps the most distinguishing feature of any mineral. Unfortunately, it is the hardest to see—in fact, it cannot be seen at all except through the eye of an X-ray machine, and to go into this aspect would lead us into discussions of physics and electronics which are beyond the scope of these introductory pages. Nevertheless, we can readily grasp the essential features of crystalline structure by looking at some of nature's own inorganic handiwork where, in favorable situations, growing minerals develop an almost perfect exterior crystal form which faithfully reflects the internal crystalline structure. Thus, many years before the discovery of X rays, mineralogists had deduced the existence of the internal crystalline structure of minerals just by close observation and measurement of the shapes and angles of the external forms. Not only that, they had calculated the possible number of crystal systems (six) and classes (thirty-two), with the result that, when X rays were discovered and applied to revealing the internal structure of minerals, vital information was developed much more rapidly and in a much more orderly way than would otherwise have been possible.

It has been one of those delightfully simplifying concepts that occasionally develop in science to discover that, despite the multitude and complexity of forms found on crystals, and despite the many facets of different shape and position that occur on these forms, all crystals can be grouped into just six crystal systems (see figure A): the isometric (or cubic), the tetragonal, the hexagonal (including the rhombohedral subsystem), the orthorhombic, the monoclinic, and the triclinic. These six systems can most readily be distinguished one from another if we picture each as possessing a set of unique axes to which all forms occurring in that system can conveniently be referred. Thus the isometric system is conceived to possess three axes, each at right angles to the other two, and all of equal length. A typical form is the cube, a six-sided form in which any one face is the equivalent of any other and is in a sense interchangeable with any other. Similarly, since all three axes are equal, they too are interchangeable.

In the tetragonal system, there are three axes, also all at right angles; but while two are of equal length, the third is either longer or shorter. By accepted convention, this unique axis, in the study of a tetragonal crystal, is placed vertically and is designated as the "c" axis. The two horizontal axes are then both designated as "a" (since they are equal). In the isometric system, all three axes are designated "a".

The hexagonal system is of particular interest since it is the only one that has more than three axes of reference. It has four; three of them, intersecting at 60-degree angles, lie in a horizontal plane; the fourth, which is either longer or shorter, is vertical and is designated "c", while the others are all "a".

In the orthorhombic system, there are three axes at right angles, but each is of different length,

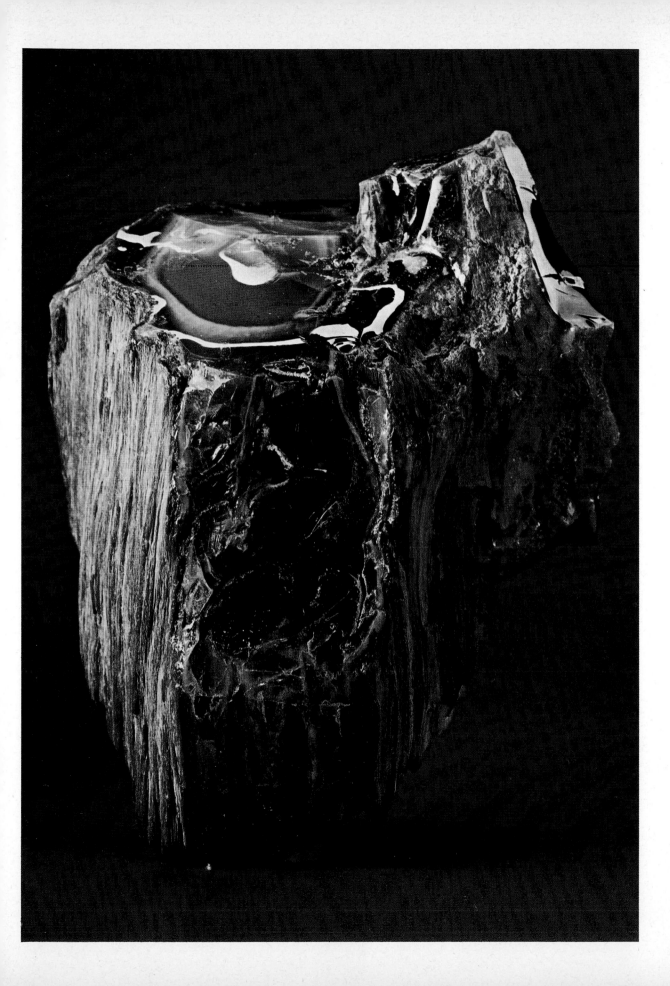

3. OPAL (petrified wood)

Collection: McGuinness
Size: 3×3 in. (8×8 cm.)
Locality: Virgin Valley, Nevada
$SiO_2 \cdot nH_2O$ Amorphous

In this replacement process of wood by inorganic silica, the exterior wood pattern has been preserved while the interior of more massive opal displays the iridescent quality that is prized by gem lovers. The name comes, with some modification, from the Sanskrit word for gem or precious stone, *upala*, indicating that men have valued this unique mineral for thousands of years.

Opal is considered amorphous, or without crystal form; that is, in amorphous substances, such as glass, the atoms have not linked up in a sufficiently well-ordered pattern to give a clear indication of crystal form.

Figure A. Drawings of crystal models illustrating the six crystal systems and their distinguishing axes of reference. All minerals crystallize within these six systems, as do all inorganic substances.

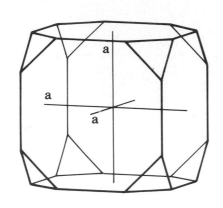

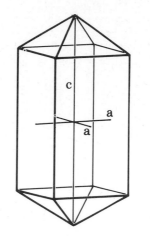

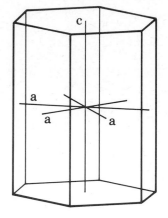

a. Isometric. Shows the three equal (therefore interchangeable) axes of the isometric system, around which are built two typical crystal forms: the dominant cube (a six-sided form) and the smaller eight-sided octahedron which modifies the corners of the cube. Such minerals as galena, pyrite, and fluorite commonly develop these forms.

b. Tetragonal. Illustrates a typical tetragonal crystal (such as zircon) in which the four-sided prism is the dominant form, terminated at each end by an eight-sided pyramid (more properly called a di-pyramid, because a pyramid shape is present both on top and bottom). The two horizontal axes "a" are of equal length and interchangeable; the vertical axis "c" is unique, and may be either longer or shorter, depending on the mineral species.

c. Hexagonal. In this hexagonal form, characteristic of such minerals as beryl and apatite, a simple six-sided prism is terminated at each end by the two faces of a basal pinacoid. The three horizontal axes "a" are of equal length and interchangeable; the vertical axis "c" is unique and is either longer or shorter, depending on the mineral species.

and convention requires that the shortest (sometimes the longest) be made "c" and held vertical. Of the two in the horizontal plane, "b" is the longer and extends from right to left; "a" is the shorter and extends to the front and back.

In the monoclinic system we find a still different arrangement. Here two axes meet at right angles: "c" (the longer of the two, and which is held vertical) and "b", which is horizontal and extends from right to left. A third axis, "a", makes a right angle with "b" but not with "c" and is so positioned that it slopes down toward the observer.

Finally, in the triclinic system, we have three axes, all of different length and all of them intersecting each other at other than a right angle. A study of figure A will clarify these relationships.

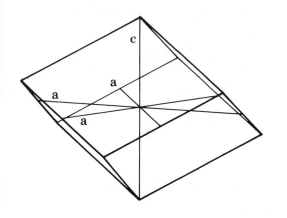

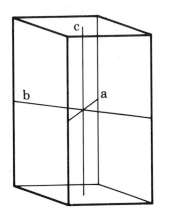

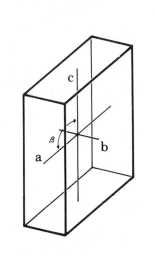

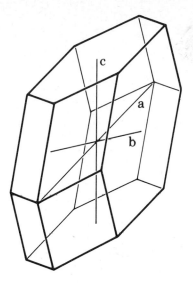

d. Hexagonal (subsystem). Illustrates the important rhombohedral subsystem of the hexagonal system. The axes of reference are the same as in *c*, but instead of a twelve-sided pyramid, there is a six-sided rhombohedron. In this form, although it is six-sided like the cube in the isometric system, all faces have the shape of a rhomb—hence the name. Rhombs are characteristic of such minerals as dolomite and siderite.

e. Orthorbombic. Shows a four-sided prism in the orthorhombic system, completed by two faces (top and bottom) of a basal pinacoid. The three axes, although all at right angles to each other, are all of different lengths.

f. Monoclinic. The monoclinic system is distinguished by the fact that although the "a" and "b" axes make a right angle with each other, as do the "b" and "c" axes, the "a" and "c" axes intersect at something other than a right angle. In the drawing, the angle *beta* is larger than 90°. The sketch shows three pinacoids: a "front pinacoid" with its two faces (front and back) parallel to the plane made by the "b" and "c" axes; a "side pinacoid" whose two faces parallel the plane made by the "a" and "c" axes; and a "basal pinacoid" whose two faces parallel the plane made by the "a" and "b" axes. Together these three two-sided pinacoid forms completely enclose space.

g. Triclinic. In the triclinic system, all three reference axes are of different lengths, and all intersect at angles other than 90°.

A word should now be said about the interesting crystal growths known as twins. In these structures, strange as they may sometimes appear, the growth patterns are governed strictly by crystallographic laws, which are not easy to describe. It will suffice here to indicate that the units of a twin are always joined in such a way that the plane of junction, or the relationship of one twin to another, can be referred to a prominent crystallographic plane (or form) or axis. And, interestingly enough, it is possible, through twinning, for triclinic crystals to simulate monoclinic crystals, and for orthorhombic crystals to simulate hexagonal crystals. Nor is twinning in crystals confined to just two units—trillings are not uncommon, and some crystals may be twinned many times over within a single specimen.

How Minerals Are Identified

Now that we know the essential characteristics of minerals, we should learn something of the way in which they can be recognized and distinguished one from another. Let us recall that chemical composition, implied in the word "homogeneous," is, along with crystalline structure, the most important determinant in a mineral. If the chemical composition of a naturally occurring homogeneous substance is accurately known, and if its crystalline structure is known, then that substance can be immediately identified and distinguished from all other substances occurring in the crust of our earth —or even on the moon!

Both properties are important. For example, calcium carbonate ($CaCO_3$) in nature commonly crystallizes in the hexagonal system and is known as the mineral calcite. But $CaCO_3$ may also, under different conditions of temperature, crystallize in the orthorhombic system, in which case it is a different mineral and is known as aragonite. Chemical analysis alone would not suffice to distinguish these two minerals. Conversely, two different minerals may exhibit identical crystal form. This is illustrated by halite ($NaCl$—common salt), which crystallizes in cubes and thus belongs to the isometric system, and galena (lead sulfide—PbS), which also crystallizes in cubes and belongs in the isometric system. Hence the cubic form in itself is insufficient to distinguish halite and galena; we must know the chemical composition as well, in order to make an accurate identification.

But since most of us who are interested in minerals, who enjoy looking at them or collecting them, do not have access to chemical laboratories and to X-ray machines we are in a sense denied the two most necessary and most accurate methods of mineral determination. Fortunately, man's ingenuity and acuity of observation have provided ready means for recognizing and differentiating many minerals on the basis of much simpler means. The most important of these are explained in the following paragraphs.

Crystal form and habit. It was mentioned earlier that observation of such characteristics as the shapes of crystals and angles between faces can often permit a highly educated guess as to the nature of the internal structure. Even more specifically, study of the crystal pattern will enable an accurate determination of the crystal system to which an unknown mineral belongs. Thus, pyrite (for instance, see plates 119–122), which is in the isometric system, can be readily distinguished from marcasite (see, for instance, plates 102–105), which belongs to the orthorhombic system despite the fact that the two minerals are identical in composition (both are FeS_2). "Habit" refers to the habit of growth of a mineral. Even if two minerals belong to the same crystal system, they may have different growth habits. Thus quartz (see, for example, plates 2 and 127–144) usually takes a prismatic habit, while calcite more commonly takes a scalenohedral habit (see, for example, plates 30 and 31). Yet both minerals are in the hexagonal system.

Specific gravity. The relative weight or heft of a mineral specimen (scientists refer to this feature as specific gravity) may give a clue to its chemical composition. Consider the difference in heft between a cube of salt, in which the sodium and the chlorine are both light elements, and galena, in which one component, lead, is a very heavy element. A figure such as "G=2.7" in a mineral description indicates that the specific gravity of such a mineral is 2.7 times that of water. This is the approximate specific gravity of many common rocks and minerals. If any mineral has a specific gravity that departs very far—either up or down—from this figure we would probably notice this as soon as we picked it up, no matter what its size, for most of us have so often picked up common rocks and stones that we sense automatically about what weight to expect from a specimen of any given size.

Hardness. Hardness, which can be revealed by scratching, or attempting to scratch a mineral, is a reflection of both composition and crystal structure, and for many minerals it provides a good first test in identification. There are curiosities here, however: for example, a diamond is the hardest of all known substances and graphite is one of the softest, yet both are pure (or nearly pure) carbon, and the chemical composition is identical. The difference in hardness in this case is wholly the result of a different arrangement of the carbon atoms in the molecule of diamond, where the atoms are rather closely packed, and in the molecule of graphite, where the packing is rather loose. This difference is further reflected by the fact that diamond crystallizes in the isometric (or cubic) system, and graphite in the hexagonal. The hardness scale used by most mineralogists, known as the Mohs' scale, is an arbitrary one based on ten minerals, each one of which will scratch all those in the scale of lower number. It runs from 1 to 10, as follows: 1 talc, 2 gypsum, 3 calcite, 4 fluorite, 5 apatite, 6 orthoclase, 7 quartz, 8 topaz, 9 corundum, 10 diamond. Thus apatite will scratch fluorite, but not orthoclase. Diamond will scratch all, and will be scratched by none. It is interesting that the absolute difference in hardness between diamond and the mineral next to it in the scale, corundum, is as great as the difference between corundum and talc.

Cleavage. Cleavage is the tendency to break smoothly or even to peel along a certain plane and is an often easily observable and rather distinctive feature of many minerals. Not only the direction and the number of cleavage planes but their quality (good, bad, or indifferent) can be a reflection of internal crystal structure. Thus the cubic cleavage (three planes all at right angles to one another and all equally good) of halite and of galena indicates conclusively that these minerals belong to the isometric crystal system, even in the absence of any cubic crystal faces. But what of a mineral that might exhibit three cleavages, all at right angles, but not of equally good quality? The answer then would be that such a mineral could not belong in the cubic system. There is, in fact, such a mineral: anhydrite (calcium sulfate, $CaSO_4$). Anhydrite is classified in the orthorhombic crystal system.

Luster. Luster is a term that somewhat vaguely calls to mind certain qualities of surface reflectance—"glossiness," "brightness," and so on—or certain kinds of very shiny pottery (lusterware). Even to mineralogists the term is a bit vague for we do not indicate luster by numbers, as we do hardness and specific gravity. But it is important, because in its broadest application it readily divides minerals into two large categories: those that possess metallic luster and those that possess nonmetallic luster.

Compare, for example, the black, shiny, almost mirror-like appearance of a cleavage surface of galena (lead sulfide, PbS) with a cube of salt or a lump of clay. The difference is obvious at first glance: the galena has a metallic luster; the salt and the clay have nonmetallic luster. This is not just a matter of color or of difference between black and white. Luster is a more fundamental quality, actually indicating something of the chemical composition and the crystal structure of a mineral.

Most minerals that contain a metal (such as silver, copper, lead, or iron) possess metallic luster. But there are many exceptions: cerussite (lead carbonate) is distinctly nonmetallic, as is hemimorphite (zinc silicate). Yet these minerals, white to transparent as they are, have a shiny appearance which distinguishes them from other white to transparent minerals such as quartz and feldspar. The former we designate as having a high luster (which results from their metal content); the latter we speak of as having low luster.

The highest luster of any nonmetallic mineral is shown by diamond, and for such an especially high luster we have a special term: adamantine. This word derives from the same Greek and Latin roots as does diamond and refers to an unbreakable or unyielding quality which ancient man (mistaking hardness for toughness) associated with diamond, the hardest of all substances. Thus, even to this day, a man who is characterized as "adamant" is considered "unbreakable" and "unyielding"—an interesting example of the influence of the terms of mineralogy on our vocabulary.

Many minerals that possess metallic luster are black or very nearly so. (Native gold, silver, and copper are important exceptions.) But not all minerals that look black possess metallic luster. Compare, for example, a specimen of shiny black tourmaline and a specimen of shiny black magnetite. They look very much alike, yet the tourmaline is nonmetallic in luster, whereas the magnetite is metallic. How then can we tell metallic and nonmetallic apart? Here perhaps is the place to mention a useful little device known as a streak plate. This is nothing more than a small plate of unglazed porcelain (a piece of broken china can make an acceptable substitute) which has a hard and somewhat roughened surface. Rub an edge of a mineral on it, and the mineral will leave a streak, which is nothing more than a fine powder of that mineral. That powder is a truer index of the luster—and the color, too—of the mineral than is the mineral's surface appearance. Most metallic minerals will yield a black streak. Conversely, a white streak indicates a nonmetallic mineral.

Streaks, however, are not all black and white. They may appear in various colors, such as red, brown, yellow, green. Try streaking a piece of shiny, black hematite (iron oxide). It will give not a black streak, but a dark-red streak. This type of hematite, therefore, does not have a *truly* metallic luster. To the luster of such minerals we apply the term "submetallic."

These are the terms, then, that are commonly used to denote different kinds of luster in minerals: metallic, submetallic, adamantine, and nonmetallic (a broad category which can be further refined by such qualifiers as low, medium, high). In many of the mineral descriptions that follow, a term for luster will be included because it is an important and a rather fundamental index to recognition of mineral species.

Index of refraction. By a certain measurement, technically referred to as the index of refraction of minerals, the mineralogist with access to a microscope has the one best and simplest method for accurate mineral identification that has been devised. It could be termed a "poor man's X ray." Sometimes denoted by n, the index of refraction is a number that relates the velocity of light in a given substance to the velocity of light in air. Thus if we partly immerse a stick in water, it appears bent—the reason being that the speed of light in water is less than in air. In most minerals, the speed of light is still further reduced, the amount depending on the chemical composition and the crystalline structure of the specific mineral. The greater the reduction of the speed of light in a mineral, the higher is the index of refraction in that mineral, and in general, the higher will be its luster. Thus diamond, with its very high (adamantine) luster, has an index of refraction of 2.42; halite, which has a rather low luster, has an index of only 1.54. Interestingly enough, if there were a mineral that had an index of refraction as low as that of water (n=1.00) it would, when immersed, become invisible!

To find the index of refraction, mineralogists have established a series of liquids with known and differing indices of refraction. For determining the index of an unknown mineral, it is only necessary to find, by immersion tests, a known liquid that matches the mineral. This is a method very commonly used by gemmologists to identify unknown stones, since it involves a completely nondestructive procedure.

Color. Other properties of minerals, though less directly reflecting fundamental features, are also useful in the determination of a mineral species. Only one will be discussed at this point: color. Although it is true that color is, quite literally, the first thing to hit the eye in any observation of a mineral and while, in gemmology, it may be one of the most valuable properties of a mineral, it should be emphasized that color is in many cases among the most superficial of mineral properties. This is nowhere better shown than in the many different colored varieties of quartz (see, for instance, plate 2). Quartz is one of the most common minerals in the crust of the earth, and it is found in a wide variety of geological (environmental, if you like) situations. But wherever found, it consists of one atom of silicon linked in a complex way with two atoms of oxygen, giving the consistent chemical composition of silicon oxide (SiO_2) and always showing in X-ray studies the same crystalline structure—which places it in the hexagonal crystal system. In fact, many quartz crystals develop faces which clearly show an elongated hexagonal growth (the "prism" faces, often capped by "pyramid" faces). Quartz is known in crystal-clear varieties (sometimes sold as "Herkimer diamonds"); in white (milky quartz); in violet to purple (amethyst); in yellow (citrine, or Spanish topaz); and in many other colors, including an almost black variety (smoky quartz or cairngorm). To the superficial observer these might well appear to be different minerals, yet to the mineralogist, the crystallographer, and the chemist they are all quartz.

Here I cannot forbear recounting an anecdote—told to me not long ago by a retired mining engineer who had been prevailed upon to give a lecture on mineralogy to a group of second-graders, largely from a ghetto area. He puzzled over what information he could convey in the one hour available to him and eventually decided that he would emphasize the importance of crystal form as a fundamental feature of minerals. Along with other illustrations, he brought a number of quartz crystals running a gamut of colors and pointed out that the angles between the faces were all exactly 60°, revealing that they all belonged to the hexagonal system, and also that the crystals were all equally hard (all would scratch glass), thereby suggesting a composition of SiO_2. At the end of the hour there were few questions and he went away wondering what, if anything, the class had learned. That weekend he met the teacher and inquired whether his talk had "gone over" at all. The teacher replied enthusiastically that it was one of the best they had ever had in a series by guest lecturers, because for days afterward the children were excitedly pointing out to each other and to all who would listen the moral they had drawn from this study of minerals: "It's not what color you are, it's what's inside that counts." Whether one proposes to espouse mineralogy or sociology, this is an important lesson to learn!

The causes of color in minerals are varied and complex. Basically they stem from the interaction of light with the electronic structures within atoms, from small "defects" in atomic and crystalline structures, and from minute amounts of impurities—that is, elements foreign to the pure, theoretical chemical composition of a mineral. A group of elements known to chemists as the transition elements is most often responsible for such coloration, and of these transition elements iron is the most abundant and the most ubiquitous. Depending on its state of oxidation and other factors, iron may produce green, blue, black, or red, yellow, or brown coloration in minerals. Other transition elements, such as manganese and chromium, also may play an important role in the color of minerals.

Even though color in some minerals can be misleading, since in many cases it is not related to the fundamental chemical composition or crystal structure of a mineral, certain colors are distinctive for certain minerals, as will be pointed out further on, in the discussions of the photographs.

It is only natural that anyone interested in minerals should want to recognize and correctly identify at least some of the more common, as well as some of the more interesting and important, minerals. By utilizing the several easily observable features just discussed and referring these to one of the many determinative tables that mineralogists have constructed, one can readily learn to make satisfactory identifications without sophisticated or expensive equipment. To substantiate this one need only recall that several hundred minerals had been described and named long before such things as analytical balances, polarizing microscopes, and X-ray spectrographs were invented.

As a further illustration, it is interesting to note that only a few years ago a hitherto very rare mineral was first correctly identified from a newly discovered locality in eastern California by an astute mining geologist using no more than an old-fashioned blowpipe (a very simple instrument with which few modern mineralogists have had any experience). The mineral was bastnaesite, a complex fluocarbonate of the rare earth metals. The deposit in which it was discovered, and correctly identified with this simple tool, is now the world's largest mine and reserve of these rare earths. Moreover, this particular bastnaesite contains a relatively high proportion of the rare earth element europium, which today gives improved color reproduction on the ubiquitous television screens. Everyone who enjoys color TV might enjoy it more if he knew the whole story of this bastnaesite discovery (recorded in the U.S. Geological Survey Professional Paper No. 261, 1954).

This Earth of Ours

Whether one's interest is in mineralogy *per se,* or whether it extends to broader matters such as mineral economics, international trade, or conservation of nonrenewable resources, sooner or later the questions must arise, "What *is* this earth of ours—what is it made of—what is it like inside—how did it get that way?"

We know that the earth is composed of three spheres: the atmosphere, the hydrosphere, and the lithosphere. The lithosphere (*lithos* in Greek means rock) equates with "solid earth," which is our present concern. We know that the solid earth is made up of rock and that rocks are composed of minerals, and we have learned that minerals are composed of distinctive combinations of elements arranged in distinctive patterns.

From astronomic and geodetic observations, we know that the solid earth is shaped not like a perfect sphere but more like one slightly flattened at the poles and slightly bulged—as a result of rotational stresses—in the equatorial area, and, from recent data from space observations, we now know that our "globe" is actually, albeit *very slightly,* pear-shaped. We know that its circumference, at the equator, is 24,902.39 miles; its area is 196,950,284 square miles; its mass, 6 sextillion, 588 quintillion short tons. From such data we also know that its density (specific gravity) is 5.51, significantly greater than the specific gravity of any of the common rocks and minerals found at the earth's surface. (For example, quartz has a specific gravity of only 2.66; feldspars, from 2.64 to 2.85; a few of the darker rock-forming minerals, such as olivine and pyroxene, may range up to 3.3.) Obviously, things must be different in the interior of the earth to account for its over-all density of 5.51.

There is an old saying among mining men that "a geologist can't see any farther into the rocks than any other man, but he can make a more educated guess as to what is down there." How does

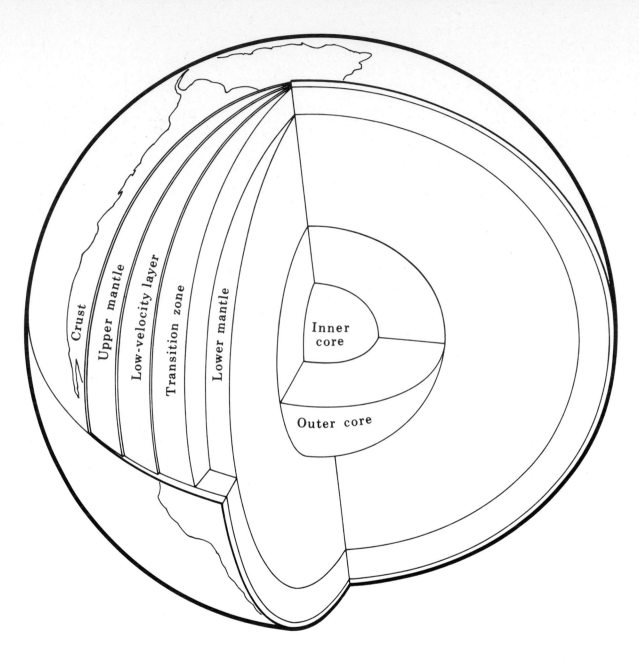

Figure B. A section showing the internal structure of the earth. (Adapted from U.S. Geological Survey Circular No. 532, *The Interior of the Earth—An Elementary Description,* by Eugene C. Robertson, 1966.)

a geologist make his educated guess? There are many ways. The first, and most obvious, is not to guess at all. He can go down to study and sample the rocks and minerals to a depth—in a few favorable spots such as deep gold mines in South Africa and Brazil—of nearly 2 miles. Almost as satisfying, in the way of direct observation, is the opportunity provided by samples of rock brought up in deep-core tests in the course of exploration for oil. The deepest of these has recently gone more than 30,050 feet below the surface, in Beckham County, Oklahoma. Yet, when we consider that it is 3963 miles to the center of the earth, even a hole 5.7 miles deep hasn't much more than scratched the surface. What about the remaining 3958 miles?

Some natural phenomena provide help to the geologist. Volcanoes, for example, which may be fed by magma generated at depths in the earth perhaps as far down as 100 miles, may spew this molten material out onto the surface where it can be examined for its chemical and mineral content. Moreover, the magma on its way up the conduit, or throat, of the volcano may tear off chunks of rock from the walls of the conduit and, in tossing them out on the surface or as inclusions in the lava, provide a somewhat random sampling of rocks at intermediate depths. Contortions that develop from time to time within the crust, such as folding, faulting, and overturning of large blocks of the crust, may bring to the surface whole sections of rock that were at one time buried at depths of many miles.

But for real penetration of the earth, into and beyond the crustal layers, we must turn to earthquakes and the records that earthquake waves place on the recording instruments known as seismographs. The seismic waves generated by a large earthquake provide a means for "X raying" the earth, inasmuch as some of the waves penetrate to great depths before being reflected (or refracted) to the surface. Some kinds of waves actually pass right through the earth; the fact that other kinds do not is evidence that the deep interior is probably molten. More specifically (see figure B), the outer core is very likely molten, whereas the inner core may be solid. Density considerations, magnetic phenomena, and meteorites all provide additional evidence about the nature of the earth's interior.

A discussion and an evaluation of such data are outside the scope of this volume. It must suffice to present, in figure B, one recent concept of our earth, as it might appear in cross section. Not all geologists would agree on the precise details, but in broad consensus figure B presents a reasonable picture of what our earth is probably like. Certainly there are, at great depths, no great treasure-filled caverns, no little green men, no lost cities to be discerned! Note the series of shells overlying the central core, each of different density and each of different bulk composition. In the outermost layer (that is, in the "crust") silicon and oxygen are the dominant elements; in the core, heavier elements, most probably iron and possibly some nickel, predominate.

How did the earth get this way? Here we enter a more speculative field, one in which whatever is written or proposed today may be outmoded by discoveries made in man's next exploration of the moon or even of Mars. On whether the earth grew by accretion of planetesimals (small solid bodies), by condensation from an ancestral and much larger sun through the spin-off of gases, by cooling from an originally molten globe, or in some other way, there is no general agreement, though each of these hypotheses has some merit. That in some manner, and at some very early time, the earth differentiated into layers of differing density, coincidentally also giving birth to the hydrosphere and atmosphere, seems well established. And so we have the pattern indicated in figure B—an abundance of light elements and compounds concentrated in the surface (continental) layers and increasing amounts of heavy elements (such as iron) and compounds toward and within the deep interior.

Although differentiation into a core and overlying shells may have been largely accomplished

4. QUARTZ geode

Collection: Halberstadt
Size: $2\frac{1}{2} \times 2$ in. (6×5 cm.)
Locality: Hibbing, Minnesota
SiO_2 Hexagonal

In this brilliantly colored and unusual geode, the somewhat stalactitic shapes formed by chalcedony have been colored by various iron oxides, ranging from the yellow of limonite to the red of hematite.

5. BARITE rose

Collection: Halpern
Size: $2\frac{1}{4} \times 2$ in. (6×5 cm.)
Locality: Noble County, Oklahoma
$BaSO_4$ Orthorhombic

Here barite crystals have grown within a sandy shale, most of which has subsequently been weathered away, leaving an aggregate with sandy surfaces—the pattern thus resembling a rose.

early in the earth's history—perhaps more than four billion years ago—it would be a mistake to regard the earth as having been a static body ever since. Far from it! Geologists have deduced many significant and dynamic episodes, the more obvious ones affecting the surface, but some of equal or greater significance affecting the earth's interior. There have been risings and fallings of sea levels; onset and advance, and retreat and disappearance of glaciations; earthquakes; volcanism; mountain-making movements; long-continued erosional forces that at times have almost leveled entire continents; internal upheavals; and breaking and sliding of continents away from and sometimes into and past each other.

Yet through all this life has persisted for at least the last three and a half billion years. Not only has it persisted, but it has been continually changing, as earlier forms have become extinct and newer forms have evolved. Paleontologists estimate that, of the many forms of life this earth has seen, more than 98 percent of the species once living are now extinct; man, who has been around for perhaps no more than two million years, is one of the 2 percent still alive today and is perhaps the only one of all these species asking such questions as why and how. Not only is he asking important questions, he is beginning to get some of the answers.

Once again the reader must be reminded that this chapter is only an introduction to a pictorial display of some outstanding mineral specimens. We cannot therefore pursue further some of the fascinating topics that a discussion of the earth brings up, but for those who may be intrigued by the story of the earth's physical evolution or by the evolution of life on the earth, references are provided at the end of the text that will tell the full story—at least as fully as it can be told today; tomorrow it may be different.

The Earth's Crust —Where Minerals Abound

Now that we know something of what our earth is like,. at least in cross section, we can direct our attention once again to minerals and ask, "Where are minerals found?" Since minerals are the basic constituents of the earth's crust, an easy first answer would be "everywhere." Looking broadly at the earth's surface, what we see is not a host of individual minerals, but rocks—rocks which are familiar to many of us: granite, basalt, limestone, sandstone, shale, and so on. Rocks are, in fact, the major constituents of the earth's crust. So, before looking for minerals, we should first know something about rocks.

Rocks are very simply classified into three great groups, based on their origins. There are igneous rocks (the word "igneous" comes from the same Greek word that gives us "ignite")—that is, those that have solidified on cooling from a previously very hot, molten condition. Geologists refer to such molten material as magma—see figures C and D. There are also sedimentary rocks, formed by the deposition of materials such as sand grains, eroded from pre-existing rocks. And there is a complex but very interesting group known as metamorphic rocks. Here, again, a Greek word is at the root; to metamorphose means to change, and a metamorphic rock is one whose original form and/or substance has been changed sufficiently so that it has taken on a distinctly different appearance from the original. Such changes can be brought about by heat, pressure, or chemical action, usually over long periods of time.

Most rocks, whether igneous, sedimentary, or metamorphic, are composed of several minerals, although a few rocks are composed of only one mineral: for example, limestone, composed wholly of calcite; or quartzite, composed wholly of quartz. So, within each of the three groups of rocks, distinctions can be based on the presence and abundance of the constituent minerals, and also upon their patterns and arrangements within the rock—commonly referred to as texture.

Hard, cold, and lifeless as most rocks are, they nevertheless can be compared—in at least one respect—to people. We know that the two major influences in the making of an individual are heredity and environment. But rocks, too, are the product of their heredity and their environment.

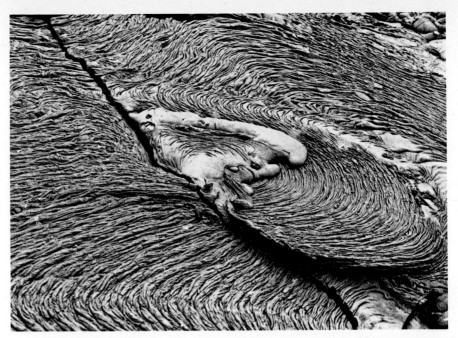

Figure C (above). This lava "fountain" and the stream of lava pouring through a narrow throat of rock give a vivid picture of the liquidity of magma. The picture was taken during the August 1969 eruption of Mauna Ulu in Hawaii. Lava fountains reached heights of 200 feet. Fully to appreciate the tremendous energy released in such eruptions, we should recall that basaltic magma has a density (weight per unit volume) three times that of water. (Photo courtesy of U.S. Geological Survey, Hawaiian Volcano Observatory)

Figure D (above right). This lava solidified so rapidly that the ropy swirls, developed as the lava stream poured from a Galápagos Island volcano in 1890, were "frozen" while still in motion, and thus still convey a sense of the liquidity of magma. Sullivan Bay, James Island, Galápagos Group. (Photo courtesy of Dr. Bruce Nolf, Central Oregon College)

Figure E (right). Lenkert Dome in Yosemite National Park conveys a sense of the massive character and coarsely crystalline texture of a typical granite. This is a small portion of a granite pluton, the name given to a large mass of magma which has slowly solidified far beneath the surface. Exhumed by the slow forces of erosion, the final sculpture, in this case of Lenkert Dome, has resulted from glaciation. (Photo courtesy of Mary Hill, California Division of Mines and Geology)

Let us look into one or two interesting examples. Granite (see figure E) is a rock familiar to most of us; so is obsidian. Anyone, at a glance, could tell these two rocks apart. Yet both are igneous rocks (having cooled and solidified from a hot silicate melt, that is, magma) and—surprising as this may seem—the heredity of both rocks is identical. Grind up a sample of each rock, give them to a chemist to analyze, and he will report the same chemical elements, in almost exactly the same amounts, in each rock; from chemical analysis alone one is unable to tell these rocks apart. Hence the very real difference in them must be a reflection of the different environments in which they were formed (that is, solidified). In the case of obsidian, the magma was poured out as lava on the earth's surface from the throat of a volcano. It cooled so rapidly that the chemical elements did not have time to get together to form recognizable crystals, and the result is a smooth, homogeneous glass—ideal, as we know, for being shaped into arrowheads and scrapers, used by early man; in World War II it was used as a very desirable substitute for man-made glass in telescope lenses. But in the case of granite, with its identical magma, solidification took place deep in the earth, perhaps several thousand feet or even some miles below the surface. Here cooling was a very slow process, and ample time was available for the formation of large crystals of the different minerals that characterize granite.

Compare next a piece of obsidian with a piece of dark basaltic lava, such as might be found around the Hawaiian volcanoes. We might have a little difficulty in telling these apart, at first glance. But on analysis a chemist could easily differentiate them, for the obsidian would show a large amount of silicon, very little iron, and probably no magnesium, whereas the basalt would show a much smaller percentage of silicon and much more iron and magnesium. In other words, we have in obsidian and basalt a pair of rocks that have very different heredity but were formed in the same environment.

For one more example, compare limestone and marble. Here are two rocks in which the heredity (that is, the chemical composition) is identical, but the marble (itself once a limestone) has been through vicissitudes of heat and pressure within the earth which have brought about a recrystallization and growth of calcite grains to produce a rock capable of taking a polish such as would be impossible for a limestone. The limestone is a sedimentary rock; the marble is a metamorphic rock, and the very name metamorphic implies an environmental episode never experienced by an ordinary limestone.

Deductions of rock history, along these lines, can readily be expanded to deductions of earth history, and they thereby provide much that is basic to the science of geology. Geology is a large field of investigation in its own right and provides an excellent foundation for the study of minerals. It has been touched on here because, without some knowledge of rocks and of geology, a search for minerals could be somewhat meaningless. And, perhaps more important, this glimpse of mineral and rock relationships (there will be more glimpses as we go along) can imbue a mineral specimen or photograph with a sense of life and dynamism.

Mineral Specimens

Even if, as mentioned earlier, minerals are found everywhere and form the bulk of the earth's crust, it does not necessarily follow that mineral *specimens* (the minerals that we like to look at, to photograph, to put in reference and in display collections) are found everywhere. Quite the contrary; good specimens of minerals are found in limited numbers and only in rather special situations within the earth's crust.

One important requirement, if a mineral is to grow to much more than microscopic size, is a sufficient supply of its chemical components furnished over a long enough time to develop a sizable crystal. There is as yet only meager knowledge of the time required for a crystal to grow, and certainly growth rates differ for different minerals and even for the same mineral, under different environments. But we can say that, for the growth of a sizable mineral crystal, tens to perhaps hundreds of years may be needed. Besides a supply of nutrients, and time, a mineral must have space in which to grow—especially if it is to develop its own characteristic crystal shape.

Consider the conditions under which silica (SiO_2) in a granite magma solidifies to form quartz. The temperature at which crystallization begins is probably about 600° C or higher. But at this temperature other minerals, such as feldspars, have already started to solidify. The magma is moderately viscous, pressures within the earth are high, and the cooling rate may be fairly rapid. The result is that, as temperatures drop to the point where most of the magma must become solidified, the molecules of silica have no opportunity to develop the well-shaped hexagonal prisms so characteristic of some specimens of quartz, because all around there are already solidified crystals of feldspar. The result: quartz is confined to whatever irregular space is left, and it forms only small, irregularly shaped grains.

But if fluids from a slowly crystallizing granite magma find their way (as they sometimes do) into cracks in overlying rocks where pressures are lower and where open spaces exist, a vein of quartz, deposited from the silica-rich fluid, may form. It may constitute a mass of milky-white quartz; it may carry within it traces of metallic gold and bring a gleam to a prospector's eye; it may develop into a mine! As miners open up the vein, places may be found where there was not enough silica to fill completely that particular portion of the crack in the crust which was the precursor of this vein of quartz. Thus there was open space in which well-shaped crystals of quartz could form—a mineral collector's dream. In other words, one of the places to look for good specimens is where open spaces have existed at the time of mineral formation and where the solutions have not completely filled the space with mineral matter.

It should be noted that, in the course of the filling of a vein by the mineral "nutrients" carried by solutions within the earth, the chemical composition, as well as the temperature, may change. The result is that a certain mineral may be deposited along the walls of a vein, but before the vein is filled another mineral—or sometimes several other minerals—may be deposited. By noting the relationships of one such mineral to another, geologists and mineralogists can work out the often complex history of a vein. This is not merely an interesting scientific exercise; it can be of considerable importance to a practical miner who needs to know (for intelligent exploration) which ore minerals tend to be early and which tend to be late in the history of vein formation. For studies of this sort, we once again have a Greek name, "paragenesis." The paragenesis of a vein, or of an ore deposit, is simply the outline of where and when (more rarely how) the minerals in a vein or an ore deposit relate to one another. If only two or three minerals are involved, as in some veins, the paragenesis is simple and easily deduced. If many minerals are involved, as is the case in complex ore deposits, the paragenesis is often difficult to work out and may be subject to considerable scientific argument.

Openings in rocks into which mineral solutions may penetrate and deposit their material occur in a variety of situations. Best known, and perhaps most productive of good specimen material are veins, discussed above. There are also solution cavities, usually found in rocks such as limestone (sometimes gypsum) which are relatively soluble in ground water. With changing conditions, there may be deposition instead of solution—even of the same mineral; we see this in the formation of stalactites and stalagmites in limestone caverns. Gas pockets may form in lava which, when the lava solidifies, may preserve the space of the bubble. Into such spaces, called amygdaloidal cavities, solutions may later precipitate many interesting minerals. These, in their paragenesis, often reflect very beautifully the declining temperature of the surrounding lava. But a physical opening is not always required for well-shaped crystals to form. It will at times suffice if the surrounding medium is a liquid rather than air. Good crystals and occasionally rare minerals may precipitate on the bottom of a lake or even on the bottom of an ocean. And some very common minerals which tend to complete their crystallization early in the cooling history of an igneous melt exhibit excellent crystal form. Olivine in basaltic rocks and feldspars in granitic rocks occasionally illustrate such crystal growth.

Some Mineral Oddities

In nature it is not the exception that proves the rule. On the contrary, nature tends to eliminate exceptions to her rule. Nevertheless, there are exceptions—or seeming exceptions—to some of the "rules" of mineralogy. And some of these are so interesting as to deserve special mention.

Pseudomorphs. One of the most striking of these exceptions involves the phenomenon of pseudomorphism. Those who know their Greek will already have figured out what this term implies for, literally translated, "pseudomorph" simply means "false form."

Perhaps the best known example of pseudomorphism is petrified wood. Specimens can be found in which the color and the texture (or grain) of the original wood are almost perfectly preserved, yet the phenomenon of pseudomorphism has changed the original organic matter (largely cellulose) of the wood to inorganic silica (SiO_2)—a chemical compound we recognize as quartz—or it may sometimes be chalcedony or opal (see plate 3). Imagine the chagrin of a woodsman who might mistakenly try to sink his axe into a chunk of this material, only to find that the woody-looking substance is now a mineral harder than the steel of his axe blade, or the surprise of a youngster gathering wood for a campfire who attempts to pick up a likely looking small log, only to find that—far from being something that would float on water—the specimen is over two and a half times heavier than water and completely unburnable!

As another example of how misleading a false form can be, look at limonite (a hydrous oxide of iron), pseudomorphous after pyrite. Every mineralogist learns very early to identify pyrite (iron sulfide, FeS_2), in part because it is perhaps the most widely distributed of all metallic minerals, in part because it is the fool's gold of the early miners—and any mineralogist worthy of the name had better be able to spot fool's gold for what it really is, whenever and wherever he sees it! It happens that pyrite often occurs in a very distinctive crystal form—named, in fact, a pyritohedron because pyrite so frequently exhibits this form, either along with or instead of the more common but much less distinctive cubic form. A pyritohedron is a twelve-sided form, in which each face has a pentagonal outline. Many mineralogists, on seeing only a single face of a pyritohedron, will identify the mineral as pyrite. And about ninety-eight times out of a hundred they will be right. But the ninety-ninth time they may be wrong because the mineral might be cobaltite (cobalt arsenic sulfide, $CoAsS$), a much rarer species, which, like pyrite, does sometimes show the pyritohedral form. And the one-hundredth time might be wrong because this particular pyritohedron is a pseudomorph of limonite after pyrite. What has happened to cause this?

Nature, certainly without intending any deception, but responsive to the laws of physical chemistry, has in this case so slowly and carefully replaced every atom of sulfur with oxygen and hydroxyl (OH) ions without significant disturbance of the iron atoms that the mineral we see is no longer pyrite (FeS_2) but $FeO(OH)$—a very different chemical compound, to which we give the name limonite.

The process of replacement, which brings about pseudomorphism, may be visualized by picturing a wall of red brick in which someone substitutes one by one white magnesia bricks for the red. Although the result will be a wall of the same size and pattern, the substance of the bricks is no longer aluminum silicate but magnesium oxide; the bricks are harder and denser, and the color is different. Now, instead of bricks, think of substitution going on, at a submicroscopic scale and without disturbance of the structure, of atoms of one element for atoms of another: the result will be a pseudomorphous mineral.

Pseudomorphs are not abundant, but they are widespread throughout the mineral world. Among mineral collectors a good pseudomorph may have added value over other specimens of the same mineral, much as to a philatelist a stamp with an inverted design has added value over the regular issue. Moreover, pseudomorphs are of considerable scientific interest, providing, as they do, incontrovertible evidence of the existence and the direction of replacement in the mineral world.

Some well-known examples of pseudomorphism, in addition to limonite after pyrite, are: quartz after calcite, quartz after fluorite, malachite after cuprite, barite after quartz, kaolinite after feldspar, chlorite after garnet, gypsum after anhydrite.

Concretions. A concretion is by no means always a pure mineral, but since it owes its origin to mineralizing processes, it deserves mention in any discussion of mineral oddities—all the more so in that few natural inorganic features, for their size, have provoked more puzzlement and led to more heated arguments, especially among laymen, than have concretions.

A concretion is simply an accumulation of mineral matter (commonly calcite, silica, or iron oxide) in the pore spaces of a sedimentary rock such as sandstone or shale. The accumulation very often starts around a nucleus, such as a shell or a leaf, from which the mineral growth outward may be either radial or concentric. In either case, if the growth takes place at an even rate, a spherical body develops, much denser and more cohesive than the surrounding unmineralized porous rock. On weathering, the spherical mass, which is more resistant, may separate out completely or may appear on a partially weathered and eroded surface, as a hemispherical projection. If of suitable dimensions—and many concretions are—they will look very much like old-fashioned cannonballs. Some rock strata have in fact been given the name cannonball formation, because of the size, abundance, and sphericity of the concretions contained.

Not all concretionary growth takes place under such ideal conditions as to yield large spheres. More commonly, the structures that develop are small discoid shapes and even more irregular patterns. Frequently, adjoining concretionary growths will partially coalesce to give a variety of weird shapes, some of which when weathered out of the rock leave patterns resembling giant footprints, handprints, tracks of imaginary animals, and so on. So realistic are some of these that laymen are often immediately convinced that the patterns were indeed made by giant men—in some cases, in very ancient strata, even as old as the Paleozoic era. Such misidentifications have been used in attempts to disprove stratigraphic sequences, to discredit geology—and even the theory of evolution.

A little careful observation on the part of geologists, mineralogists, or laymen will usually reveal the real nature of these structures. That they represent growth features is shown by the fact that the sedimentary bands of the original rock can often be seen to pass right through the concretion. In other words, the new mineral matter was so gently accumulated around the nucleus that the growth did not disturb the sedimentary layers. When such growth takes place—as it often does—as a series of concentric bands, the development of a concretion can in a crude way be likened to the growth of a pearl around a sand grain in an oyster. Both the pearl and the concretion possess an onionskin structure; at the center of the pearl and often at the center of a concretion will be a foreign body.

But there our comparison must end. Unlike pearls, which no one would think of breaking open in order to find a sand grain, breaking open a concretion may reveal a well-preserved and interesting fossil.

Geodes. Geodes are rough-surfaced bodies, subspherical in shape, which range in size from an inch or two to sometimes as much as a foot or more in diameter (see plate 4). In external appearance, except for their usually much rougher outer surface, they are not unlike concretions. But they differ from concretions in being hollow and in that they most commonly occur in limestone (occasionally in shale). Moreover, the outermost rind is invariably chalcedony (fine-grained silica, rather similar to quartz). When broken open, crystals of quartz and/or calcite, and sometimes a variety of other minerals are found projecting inward into the internal cavity. Because of the open space into which these minerals grew, they commonly have developed excellent crystal faces and terminations. Still another difference from concretions is that many geodes appear to have expanded during growth, having forcibly pushed aside surrounding strata in order to make room for themselves, unlike concretions which grow, as it were, passively. The actual growth mechanism of geodes cannot be simply explained. Essentially it involves deposition of initially gelatinous silica which acts as an osmotic membrane to build pressure and selectively to concentrate the chemical constituents found on the inside. These may include the ingredients for such minerals as pyrite (iron sulfide, FeS_2) and even millerite (nickel sulfide, NiS). In any event geodes are eagerly sought by collectors because of the interesting contrast between the rough, unlovely, chalcedonic exterior surface and the glittering, many faceted, well-formed crystals on the inside.

Septaria. Septaria are large nodular structures that range up to two or three feet in largest dimension and are characterized by radiating cracks widening toward the center. The radiating cracks may be transected by somewhat concentric cracks—but all, radials and concentrics, are usually extremely irregular. Most commonly the cracks are filled with calcite (see plate 6) which, being coarser-grained and more resistant, yields a curious pattern of ridges on a weathered surface—sometimes providing a fanciful imitation of a turtle shell! Septarian nodules are usually found in calcareous shales. Their origin involves formation, initially, of an alumina gel, with later concentration and veining of calcite. Although formed by inorganic processes, septarian nodules are occasionally found to have fossil cores. Particularly spectacular are nodules built around coiled ammonites.

Dendrites. Of the many ways in which inorganic nature may seem to simulate life, dendrites are among the best examples. The name itself comes from the Greek word *dendro*, meaning a tree (from which we have the word dendrology, the science of trees).

A mineral dendrite is nevertheless a strictly inorganic growth in which a crystal pattern develops into an arborescent or branching texture—usually in the course of evaporation of a miner-

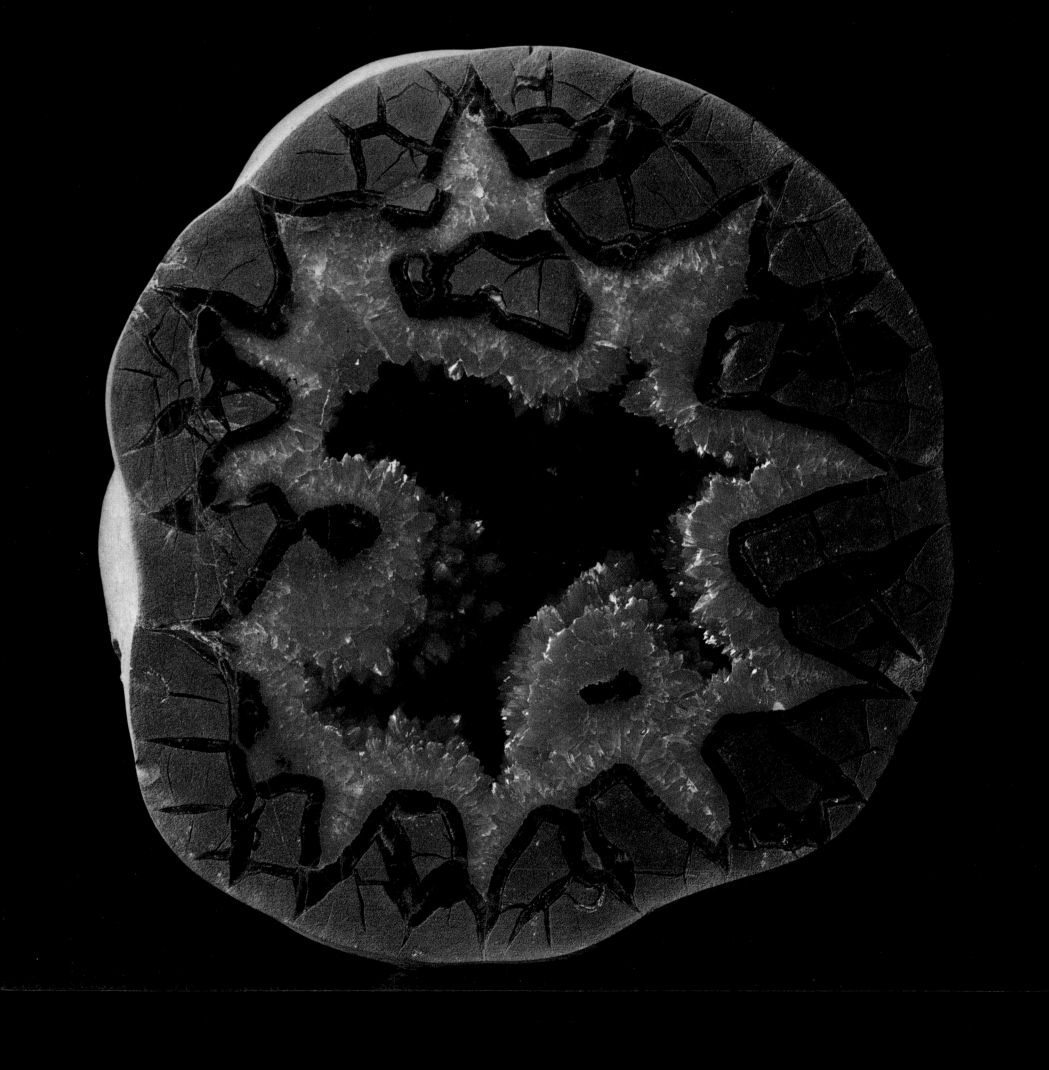

6. CALCITE (a septarian nodule)

Collection: Court
Size: 7×7 in. (18×18 cm.)
Locality: Orderville, Utah
$CaCO_3$ Hexagonal

In this cut and polished section through the center of a septarian nodule masses of calcite crystals have not only lined all the walls of the internal cavity but have almost completely filled it.

alizing solution that has penetrated along thin bedding planes of a sedimentary rock. When a slab of such a rock is broken open along a bedding plane, a structure is revealed which looks like a fossilized fern or leaf frond. Most dendrites are black and are composed of manganese oxides; some are brown or red and in such cases may be oxides of iron.

Well developed dendrites make attractive display specimens and can serve to remind us that things are not always what they may seem, at first glance, to be!

Pegmatites. Strictly speaking, we must consider pegmatites as rocks. But because pegmatites are *the* happy hunting ground for minerals, whether it be for rare species, for well-formed crystals, for gem-quality material, or for giant-sized specimens, these interesting structures surely deserve special mention.

Simply defined, a pegmatite is an exceptionally coarse-grained igneous rock. Granite pegmatites are by far the most abundant and are of most interest to mineralogists. Nevertheless, pegmatitic varieties of other plutonic rocks are known. In shape, pegmatites range from tabular and lens-like to highly irregular. In size, they may be little more than a small vein or they may be tens of feet thick and hundreds of feet long. Most granite pegmatites form in or near the borders of large granitic intrusives. The minerals that occur most abundantly, and often the only minerals to be found in a granite pegmatite, are quartz and feldspar. Mica also may be a common constituent. Individual crystals are commonly measured in inches, but may range up to several tens of feet. Pegmatites may contain many rare minerals, some occasionally of gem quality. And since some pegmatites contain internal open spaces, well-formed crystals within the interior of a pegmatite are not uncommon. Such "pegmatite pockets" are treasure troves indeed, for these are the source of much gem-quality tourmaline, kunzite, beryl, and other minerals.

The formation of a pegmatite is the end stage in the consolidation-crystallization history of a magma that has cooled slowly at some considerable depth in the earth's crust. As a plutonic magma cools, first the high-temperature minerals crystallize, then the lower-temperature minerals gradually crystallize. Toward the end of this process the fluid magma that remains may have become enriched in rare elements and gases which lower viscosity and promote mobility of the fluid. This residual "juice" may be squeezed out into the country rock adjoining the major intrusive, or it may form a fluid body within the still warm but essentially solidified parent. Depending on further conditions, some or all of the gases may escape, leaving only quartz and feldspar to form the pegmatite, or all of the gases may be trapped. Particularly in the latter case, the lowering of viscosity permits crystals to grow to exceptional size. A single crystal of spodumene (lithium-aluminum-silicate) forty-seven feet long was mined for several years for its lithium content in a pegmatite in the Black Hills of South Dakota. Sheets of mica, several feet across, have been mined in some pegmatites in Canada, and crystals of quartz and feldspar many feet long are known from numerous locations.

Although crystal size is the most obvious feature of pegmatites, equally distinctive is the concentration of otherwise rare elements such as lithium, beryllium, boron, and phosphorus that may occur in the more complex pegmatites and that lead, on crystallization, to the many unusual minerals for which pegmatites are famous.

Envoi

Those who have followed this commentary thus far are now about to start on a scenic journey—one that constitutes the main part of this volume. On this journey we will see a wide variety of minerals: rare minerals, common minerals, minerals vital to man, minerals of little or no use to man. Yet all are attractive specimens; as a distinguished scientist recently remarked in the course of his erudite presidential address before the Mineralogical Society of America, "the best thing about minerals is the way they look."

But those who have followed these pages know that there is often much more to a mineral than just the way it looks. Therefore, in the next section, along with the photographs, some discussion will be included.

For those who, following this introduction to the mineral kingdom, want to explore it further, there is a short list of books from which much additional information, as well as further references, can readily be gained.

Besides reading about minerals, viewing is highly recommended. No one interested in minerals should miss any opportunity to see the many fine mineral displays now in museums throughout the world. Perhaps the finest of all collections—especially for wealth of material—is in the British Museum of Natural History in South Kensington, London. Paris, with the displays at the Sorbonne and at the École des Mines, has two excellent mineral museums, and also other European capitals have collections well worth a traveler's time. Even the hastiest glance across North America would take note of such collections as those in the Hall of Minerals at the California Academy of Sciences in San Francisco; the Field Museum of Natural History in Chicago; the Cranbrook Institute of Science in Bloomfield Hills, Michigan; the Philadelphia Academy of Sciences; the Royal Ontario Museum in Toronto; the Canadian National Museum in Ottawa; Harvard University in Cambridge, Massachusetts; and the American Museum of Natural History in New York City. Preeminent, of course, are the collections—especially of gems—in the United States National Museum in Washington, D.C. But no one should overlook the fact that surprisingly fine collections can be discovered in unsuspected places. For example, an excellent display of native gold can be seen in the Siskiyou County Courthouse in the little town of Yreka (population 4,759) in northern California.

Finally, for those who discover a growing interest in minerals and who may be thinking of starting a collection of their own, there is no better way to get started than by getting in touch with a local gem and mineral society (visitors are almost invariably welcome). There are now thousands of these clubs throughout the country, not only in the larger cities, but in smaller towns as well.

PLATES AND COMMENTARIES

The arrangement in the following part is alphabetic, by mineral names. This is not the way in which minerals are ordinarily arranged in a textbook or a reference book, where they are presented in natural groups depending on their chemical structure (such as elements, sulfides, oxides, carbonates, silicates). But this book is not a textbook or a reference book on minerals, nor is it even a systematic sampling of minerals; it is a "browsing book." And whether we are browsers or more serious students, most of us remember objects, or people, best by their names A good precedent for a purely alphabetic arrangement of minerals is the widely used Minerals Year Book *of the U.S. Bureau of Mines.*

The description of each specimen photographed provides its name, its present owner (see list of collections in the acknowledgments), its dimensions in inches and centimeters (width first, height next), the locality where it was found, its chemical formula, and the crystal system to which it belongs. (The chemical composition is given in "chemist's shorthand." For those unfamiliar with the designations—such as K for potassium, Na for sodium—these symbols with their equivalent names are tabulated alphabetically in the appendix.) In addition, there may be a discussion pertinent to the specimen, covering such details as its color, characteristic growth habits, usefulness to man, and name. In those instances where several specimens of the same species are shown, most of the general discussion will be found under one specimen.

7. ACTINOLITE

Collection: California Academy of Sciences
Size: $11\frac{1}{2} \times 4$ in. (29×10 cm.)
Locality: Mendocino County, California
$Ca_2(Mg,Fe)_5Si_8O_{22}(OH)_2$ Monoclinic

Actinolite is a member of the important and wide-ranging amphibole family of rock-forming minerals. Actinolite is found chiefly in metamorphic rocks, unlike some better known members of this family, which occur mainly in igneous rocks. The name comes from the Greek *aktis* (ray) and *lithos* (stone), in reference to the radiating pattern in which the needle-like crystals frequently develop.

8. ADAMITE on LIMONITE

Collection: Court
Size: 15 × 10 in. (38 × 25 cm.)
Locality: Ojuela Mine, Mapimí, Durango, Mexico
Adamite: Zn_2AsO_4OH Orthorhombic
Limonite (goethite): $FeO(OH)$ Orthorhombic

Although classified as a nonmetallic mineral, because of its zinc content adamite possesses a very high, in fact an adamantine luster (see Introduction) which, along with the delicate greenish-yellow color and the dark-brown background provided by the iron oxide minerals on which the adamite crystals have grown, makes specimens like this one of a collector's most prized and spectacular exhibits. The name honors Gilbert Joseph Adam (1795–1881), a French mineralogist who supplied the first specimen for study.

9. APATITE

Collection: Court
Size: 4×6 in. (10×15 cm.)
Locality: Panasqueira, Portugal
$Ca_5(PO_4)_3(F,Cl,OH)$ Hexagonal

Apatite, in its several varieties—like quartz in its several varieties—is an outstanding example of a mineral that combines beauty, important values for man, and much scientific interest.

Although occasionally the more transparent and the more highly colored varieties of apatite are utilized as gemstones, it is in the displays of crystal growths in a wide variety of patterns and colors that apatite best demonstrates its aesthetic appeal. To the scientist, especially to the chemist and the crystallographer, the apatite crystal structure has long been of special interest because not only is substitution of major cations (Pb for Ca, for example) possible without serious disturbance of the structure, as well as substitution of minor components one for another (F for OH or Cl and vice versa), but it is also possible to substitute for the major anion complex, for example (AsO_4) or (VO_4) for (PO_4).

Interesting and important as these matters are, they pale into insignificance when the role of apatite in man and in man's modern world is considered. *No other mineral is so personally involved in man*—if we may put it that way—as is apatite, for no other mineral is so much a part of him. Most of our teeth and much of our bones are apatite or an apatite-like "mineral." To be sure, since a mineral is defined as being of inorganic origin, we cannot properly say that a thigh bone or a wisdom tooth is mostly apatite. But the fact is that the hard material of tooth or bone is virtually identical, in chemistry and structure, to apatite. Incidentally, the presence of "apatite" in skeletal structure is characteristic of (although not confined to) vertebrates and is thus, in a sense, a culmination of evolutionary developments, starting with the secretion of silica and lime by single-cell organisms millions of years before vertebrates appeared on the earth. And the element phosphorus, for which apatite is the principal source, is an essential component of every living cell, plant and animal.

Because of this vital requirement on the part of all living things for this key element, it is of interest and importance to know something of how and where it is derived. Virtually all of the phosphorus in the world today has either come from or is locked up in this one mineral, apatite. Although it rarely constitutes more than a fraction of 1 percent of most primary rocks, apatite is nevertheless very widely distributed. It is found in volcanic rocks, such as basalt; in plutonic rocks, such a granite; and in almost all intermediate types. Almost always it occurs in tiny but well shaped hexagonal prisms—an indication that no matter what type of parent magma it came from, it was invariably one of the very earliest minerals to crystallize.

So, whenever any of these rocks is exposed at the surface of the earth to weathering and decomposition, some phosphorus is released and finds its way into ground-water and stream circulation, and into soils. Since plants require phosphorus, their roots extract it from the soil and build it into their cellular structures. Then the plant material is eaten by herbivores and omnivores, and in this way phosphorus enters their cells as the element phosphorus and, in combination with calcium, as the apatite-like material of their skeletal parts.

In the meantime, of course, some of the phosphorus leached and dissolved from the primary rocks finds its way to the sea. Here fish extract it and build it into their bones; birds eat the fish, utilizing some of the phosphorus for their bones and excreting more—in some cases in such large amounts as to build up the so-called guano islands, which were at one time among the richest sources of the phosphatic fertilizers required by man to replenish his farm lands which, because of constant cropping, become deficient in the phosphorus so vitally needed by plant crops.

We have only sketched, in brief and partial outline, the phosphorus cycle—one of the more fascinating and certainly one of the most important geochemical cycles in nature. Even so, the importance of having sufficient amounts of this vital fertilizer, phosphorus, to provide food for the earth's growing population, will be clear to all. And, in this respect, nature has indeed been generous. As a result of more complicated patterns in parts of the phosphorus cycle there are in the world some tremendous accumulations of *relatively* concentrated apatite-bearing rocks—chiefly in a few sedimentary formations. But, as with other products of her mineral largesse, nature has not made an equitable distribution of apatite concentrations. In the United States, for example, minable phosphate deposits are found chiefly in Florida; in Tennessee; and in some of the northern Rocky Mountain states, Montana, Idaho, Wyoming, and Utah. In contrast, some of our richest agricultural states, such as Iowa, Illinois, Kansas, Texas, California, are virtually without phosphate deposits. The same situation exists worldwide. Outside the United States, some of the richest phosphate deposits are in North Africa (chiefly in Morocco, Algeria, and Tunisia), while most of Europe lacks phosphate deposits. There are others (some guano islands) in the South Pacific, but Japan and China are without or with only very limited deposits, and India is highly dependent on importations. In short, a few fortunate countries are "haves"; many more are "have-nots."

The mineral's name, apatite, comes from the Greek *apate* (deceit), because apatite has so often been mistaken for other minerals. It has, in fact, been known as "the deceptive stone."

10. APOPHYLLITE on PREHNITE

Collection: Land
Size: 10×5 in. $(26 \times 13$ cm.)
Locality: Paterson, New Jersey
Apophyllite: $KCa_4Si_8O_{20}(F,OH) \cdot H_2O$ Tetragonal
Prehnite: $Ca_2Al_2Si_3O_{10}(OH)_2$ Orthorhombic

Perched on a mass of small, greenish-white prehnite crystals are transparent to white tabular crystals of apophyllite, a mineral with one
excellent cleavage and an exceptionally large water content. When
fragments of apophyllite are heated in front of a blowpipe, these features combine to give the mineral a low fusibility and a tendency to
expand into leaflike forms many times the original volume. It is from
this that the mineral gets its name: from the Greek roots *apo* (from) and
phyllite (leaf). Prehnite takes its name from Colonel Prehn, governor of
Cape Colony, South Africa, who in the eighteenth century first called
attention to large masses of this attractive mineral occurring in the area.

11. APOPHYLLITE, STILBITE, and SCOLECITE

Collection: Land
Size: $5\frac{1}{2} \times 6$ in. (14×15 cm.)
Locality: Fernaneil, Goiás, Brazil
Apophyllite: $KCa_4Si_8O_{20}(F,OH) \cdot H_2O$ Tetragonal
Stilbite: $NaCa_2(Al_5Si_{13})O_{36} \cdot 14H_2O$ Monoclinic
Scolecite: $Ca(Al_2Si_3)O_{10} \cdot 3H_2O$ Monoclinic

Here, in a large cavity that was originally a gas bubble in basaltic lava, small glassy crystals of apophyllite rest on yellow stilbite, adjoining a bundle of radiating fibers of white scolecite. The smaller cavities in the basalt are called vesicles. These are lined with apophyllite and some zeolites.

12. ARAGONITE

Collection: Court
Size: 4×6 in. (10×15 cm.)
Locality: Chihuahua, Mexico
$CaCO_3$ Orthorhombic

In its chemical composition this mineral is identical with calcite but, since calcite crystallizes in the hexagonal system and aragonite in the orthorhombic, they are distinctly different minerals and have among other things different hardness and different specific gravity. When a chemical compound, such as in this instance calcium carbonate, can crystallize in either of two different systems, it is referred to as being dimorphous (having two forms). Dimorphous compounds are not uncommon among minerals. There are even some compounds that can crystallize in more than two crystal systems. Such minerals are known as polymorphs.

Much research has gone into the thermodynamic and biologic controls that determine whether aragonite or calcite will be precipitated in a given situation. As a result some "environmental" deductions can be made from occurrences of these two minerals. In general, calcite is the more stable of the dimorphs under atmospheric conditions, and it is therefore the more common of the two minerals in and on the crust of the earth. From water solutions of $CaCO_3$, containing CO_2, calcite is the usual precipitate at ordinary temperature. But at higher temperatures (sometimes also in the presence of other salts) aragonite will precipitate. In many fossil shells, a nacreous layer of aragonite seems to have formed the original precipitate by the lime-secreting animal (such as a clam) but in most instances this has inverted later to calcite. Pearl is itself composed of aragonite. The somewhat unusual growth pattern exhibited in the photograph is rather characteristic of some forms of aragonite, referred to as *flos ferri* (flower of iron), although iron is not an essential component. The mineral name comes from the province of Aragon in Spain.

13. ARAGONITE

Collection: Land
Size: $1\frac{1}{4} \times 1\frac{1}{4}$ in. $(3 \times 3$ cm.)
Locality: Aragon, Spain
$CaCO_3$ Orthorhombic

Look at the dark central portion of this very interesting specimen. At first sight it might be a cube. Now turn the picture 90 degrees and look carefully: what might have been a cube is seen to be a hexagonal prism, with a flat top and base. But aragonite is not hexagonal; it's an orthorhombic mineral! The explanation lies in the fact that aragonite frequently forms not twins but trillings (see the Introduction). That is, these individuals interpenetrate at exactly 120-degree angles, thereby yielding a "pseudo-hexagonal" whole. Around the original trilling, another, somewhat lighter-colored, trilling has grown to make a most unusual specimen.

14. ARSENOPYRITE with WOLFRAMITE and
     QUARTZ

Collection: Court
Size: $4\frac{1}{2} \times 4$ in. (11×10 cm.)
Locality: Panasqueira, Portugal
Arsenopyrite: FeAsS   Orthorhombic
Wolframite: (Fe,Mn)WO$_4$ Monoclinic
Quartz: SiO$_2$ Hexagonal

The brassy-colored mineral is arsenopyrite; wolframite is the shiny
black submetallic mineral which here displays to advantage its excellent
pinacoidal cleavage. (Wolframite, incidentally, is one of the two principal ore minerals of tungsten.) Well shaped, glassy, hexagonal crystals of
quartz are seen projecting from the mass of arsenopyrite and wolframite. The name arsenopyrite comes simply from a combination of arsenic
plus pyrite, indicating its resemblance (in color and hardness) to pyrite
and its content of arsenic.

15. AUTUNITE

Collection: Land
Size: $4 \times 5\frac{1}{2}$ in. (10×14 cm.)
Locality: Daybreak Mine, Mount Spokane,
             Washington

$Ca(UO_2)_2(PO_4)_2 \cdot 10-12H_2O$ Tetragonal

Autunite is one of the many secondary uranium minerals characterized by their brilliant colors, usually greens and yellows. It forms as the result of breakdown of primary uranium minerals and recombination with volatile components such as phosphorus and water, as may happen in pegmatites. Spectacular specimens such as this one have been found in the Daybreak Mine near Spokane—one of the important but less well known uranium occurrences in the United States. The name comes from Autun, France, another notable locality for this mineral.

16. AXINITE

Collection: McGuinness
Size: $3\frac{1}{2} \times 3$ in. $(9 \times 8$ cm.$)$
Locality: Obira, Bungo, Japan
$(Ca,Mn,Fe)_3Al_2(BO_3)Si_4O_{12}(OH)$ Triclinic

The name first given to this mineral (in 1792 by J. C. Delametherie) was yanolite, meaning "violet stone," because many crystals do show a distinctive violet color. But a mineral in which the ratios of various chemical components, such as iron and manganese, can vary as much as is indicated in the chemical formula will obviously occur in a variety of colors. In fact, the most characteristic color for axinite is a sort of clove brown. Hence, a name proposed in 1799 by R. J. Hauy (who himself has had a distinctive mineral, hauynite, named after him) is now generally accepted—based upon the sharp-edged, wedgelike crystal form resembling an axe (from the Greek *axine,* an axe). Because of the hardness (almost 7) and the attractive colors, transparent varieties of axinite are recognized as semiprecious gemstones.

17. AZURITE with MALACHITE and LIMONITE

Collection: Court
Size: 5×7 in. (13×18 cm.)
Locality: Copper Queen Mine, Bisbee, Arizona
Azurite: $Cu_3(CO_3)_2(OH)_2$ Monoclinic
Malachite: $Cu_2(CO_3)(OH)_2$ Monoclinic
Limonite (goethite): $FeO(OH)$ Orthorhombic

In this specimen the growth pattern is botryoidal (that is, grapelike).
The rounded structures were probably formed first as azurite, which
has subsequently partially altered to malachite. (For more on malachite, see especially plate 18.)

18. AZURITE and MALACHITE

Collection: Brown
Size: $3\frac{1}{2} \times 4$ in. (9×10 cm.)
Locality: Apex Mine, St. George, Utah
Malachite: $Cu_2(CO_3)(OH)_2$ Monoclinic
Azurite: $Cu_3(CO_3)_2(OH)_2$ Monoclinic

Here we have an example of malachite and azurite casts after selenite. In this specimen it is of particular interest to note that, around the holes from which the selenite is now gone, azurite had formed first, to be followed (and partly replaced) by malachite. In fact, these two minerals are so nearly alike that very slight changes in the chemistry of the depositing solutions or even of temperature and moisture in the atmosphere can cause malachite to change to azurite, or vice versa.

Malachite is harder than it looks, harder than its delicate fibers suggest. It is in fact slightly harder than marble and, like marble, many specimens of malachite will take a high polish which accentuates the interesting color variations and textures inherent in the mineral. These features, when well displayed in sizable specimens, have made malachite a rather valuable ornamental stone.

Unfortunately, the amounts of malachite that occur in large enough masses of proper quality to yield sizable polished slabs are severely limited and most of the best of them were discovered and mined in Russia (from the Siberian, or eastern, side of the Ural Mountains) one hundred to one hundred fifty years ago. The very best specimens became the property of the czars and were used as decorative material in everything from small snuff boxes to large door panels and table tops. Many of these are still on exhibit in Russian museums and some have found their way to western Europe and America.

Although large, polishable slabs are rare, malachite as a mineral and as an attractive specimen material is found in many parts of the world. Bisbee, Arizona, is one of the best known sources in the United States. Malachite is also found throughout the Andean copper belt in western South America; in South Australia; in South-West Africa; and indeed wherever copper mineralization and carbonate rocks (such as limestone) are found together in the crust of the earth.

Malachite was a mineral known to the ancients as a source of copper. Its green color—which along with some other green copper minerals is known to prospectors as copper stain—has long been one of the guides to copper ore. The distinctive copper green, or copper stain, is easy to spot on otherwise dull rocks, even by the novice prospector. Unfortunately for many eager ore seekers, it takes only a trace of copper carbonate or copper silicate to produce an eye-catching "stain" on rocks. Hence, the presence of malachite in a thin crust by no means indicates that undiscovered copper riches lie in the rocks below. But it is an indicator that at the very least calls for a closer look and perhaps for some serious exploration and sampling of the rocks where it is found.

Malachite forms as a secondary mineral and thus seldom extends to any considerable depth below the surface. Today it is a rather minor ore of copper. But it has been important in the past, and many of the great copper mines—such as Bisbee in Arizona, Tsumeb in South-West Africa, and others—got their start as copper producers through the mining and refining of the easily worked and accessible surface and near-surface malachite.

Malachite has one very near relative in the mineral world, also a basic copper carbonate named for its color: azurite (the name refers to the "azure blue" of heaven). Not infrequently, as in the specimen shown here, the two minerals occur together, yielding attractive patterns of interbanded colors—various shades of green with deep blue. Malachite, under surface conditions, is somewhat more stable, chemically, than is azurite. Thus azurite may be found in part changed over to malachite. In fact, malachite may even assume the form of an azurite crystal through a process known as replacement, thus developing what is known to mineralogists as a pseudomorph (that is, "false form"). In other words, the crystal has the shape of azurite, but the chemical composition is that of malachite. Petrified wood is a well-known example of replacement—in this case the organic substance of the wood is replaced by inorganic silica which nevertheless faithfully preserves the grain and texture, and sometimes even the color, of the original log.

The change of azurite to malachite is a slow and subtle process that may lead to surprising results. The story is told that—back in medieval times—painters depended on finely ground lapis lazuli, a mineral of deep-blue color, to provide when mixed in oil the ultramarine blue they needed for skies and seascapes. But in medieval times, even as now, there were, alas, sharp and dishonest dealers. They learned that azurite, when finely ground, yielded practically the same shades of blue that were obtained from genuine lapis lazuli. Since azurite was much easier to come by and much cheaper, and since when it was ground in oil few if any painters could tell the difference, a lot of spurious ultramarine pigment, based on azurite, went into the market. As the story goes, even the great Titian was at times a victim of dealers who supplied azurite rather than lapis-lazuli blue. Nature's replacement of azurite by malachite is a very slow process, so evidence of this chicanery did not become apparent until long afterward, but over decades, and indeed centuries, slowly the azure blue of these skies faded to malachite green. Elsewhere, malachite has played a part in religious symbolism. For example, some scholars have held that malachite was the official decorative stone in the breastplate of the high priest of the early Hebrews.

Much more could be said about malachite, its history, its many occurrences, its mineralogical relationships, its museum displays. But the points covered should suffice to assure malachite a well-deserved place in any catalogue of the world's more interesting and attractive minerals.

The name derives from the Greek *molokhitis,* "the mallow-green stone."

19. AZURITE and MALACHITE

Collection: Brown
Size: 2×7 in. $(5 \times 18$ cm.)
Locality: Apex Mine, St. George, Utah
Azurite: $Cu_3(CO_3)_2(OH)_2$ Monoclinic
Malachite: $Cu_2(CO_3)(OH)_2$ Monoclinic

These stalagmitic growths are further evidence of the versatility of nature's mineral displays. Here we have casts of malachite and azurite after selenite—which now appear like fanciful and brightly uniformed and decorated sentries.

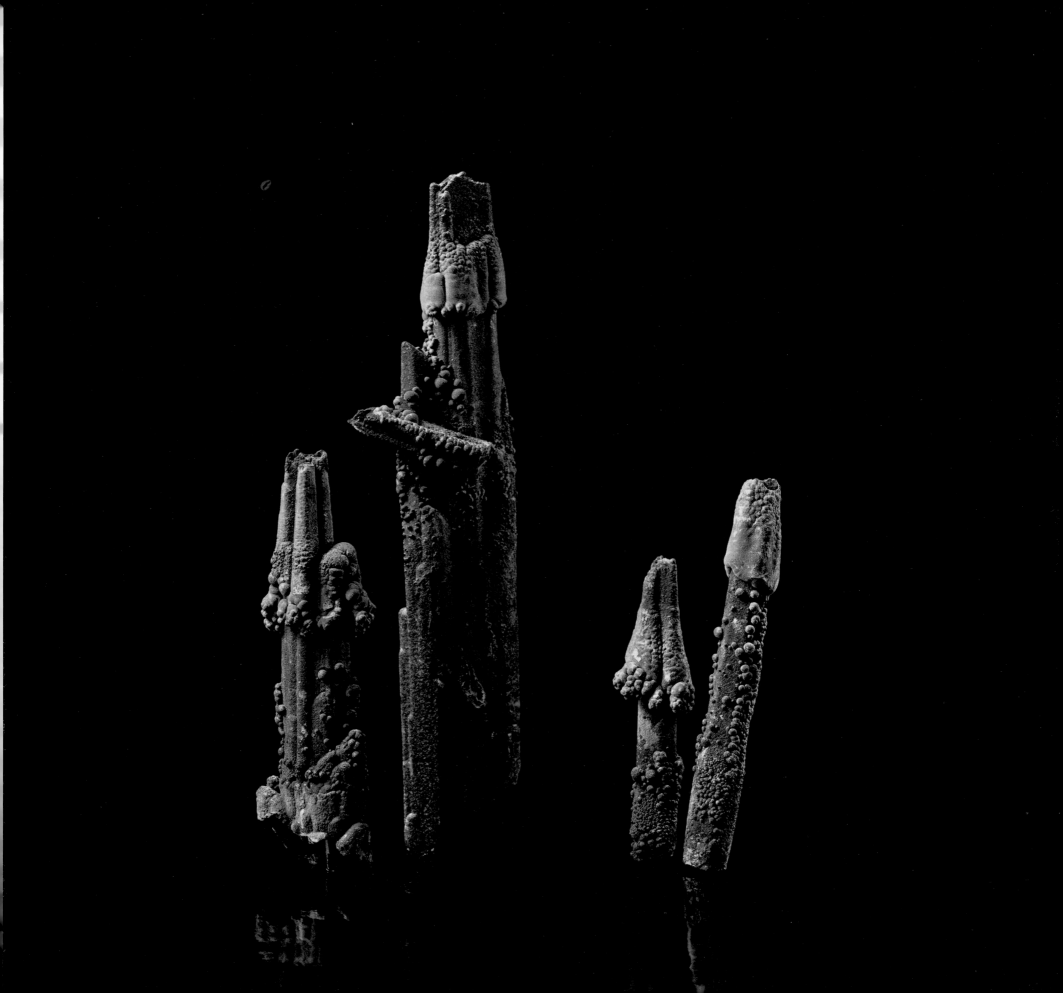

22. BARITE

Collection: Smale
Size: 4×6 in. (10×15 cm.)
Locality: Cumberland, England
$BaSO_4$ Orthorhombic

This specimen is unusual, not only for the exceptionally well-formed crystal terminations but also for their transparency and delicate golden color. Among many in the mining fraternity, barite is commonly known as heavy spar, in reference to its relatively high specific gravity ($G = 4.5$) for a nonmetallic mineral; its name comes from the Greek word *barus* (heavy). Since most white or transparent minerals of nonmetallic luster (see the Introduction) possess a specific gravity of around 2.7 (quartz is 2.65, the feldspars range from 2.57 to 2.77, calcite is 2.71, and these minerals account for perhaps nearly 90 percent of the light-colored rocks and minerals we see in the crust of the earth), when we pick up a specimen of barite, the exceptionally high specific gravity is immediately noticed. This property is not only one of the readiest ways for recognizing the mineral, even when massive and lacking visible crystal form, but is also a property which gives the mineral important industrial application. Barite is not a hard mineral ($H = 3.5$), has good cleavage, and is easily ground to a fine powder. It is also chemically very inert. These features combine to make barite indispensable in the drilling of oil wells, for which finely ground barite is mixed with water (because it is so finely ground it will remain in suspension for some time), and sometimes also with clay, to form a slurry which not only lubricates the bit as it churns its way through rock strata thousands of feet below the surface, but also provides sufficient weight in the hole so that an encounter with gas or oil under high natural pressures within the earth will not generate a blowout.

Barite at times has been used in nefarious ways, because it is easily ground to a fine powder, because of its white color, and because—being chemically inert—it is tasteless and harmless if taken internally. Dishonest dealers have been known to add it, in small amounts, to flour and to sugar, thereby increasing the selling weight while using a cheaper substance. Nevertheless, barite has its legitimate internal uses, again because of the properties mentioned. Thus a physician may have an ailing patient drink a slurry of barite, since its chemical inertness and a marked opacity to X rays permit charting the digestive system without recourse to surgical procedures. Truly, barite is a very useful mineral.

23. BARITE (golden barite) on CALCITE

Collection: Land
Size: 4×5 in. (10×13 cm.)
Locality: Elk Creek, Meade County, South Dakota
Barite: $BaSO_4$ Orthorhombic
Calcite: $CaCO_3$ Hexagonal

24. BARITE

Collection: Court
Size: 5×6 in. (13×15 cm.)
Locality: Orderville, Utah
BaSO$_4$ Orthorhombic

In this unusual specimen, markedly elongated crystals of barite have grown within a septarian nodule (see the Introduction).

25. BAYLDONITE

Collection: Krueger
Size: 2×3 in. (5×8 cm.)
Locality: Tsumeb, South-West Africa
$PbCu_3(AsO_4)_2(OH)_2$ Monoclinic

This brilliant green mineral was named for Dr. John Bayldon.

26. BENITOITE

Collection: McGuinness
Size: 4×6 in. (10×15 cm.)
Locality: San Benito County, California
$BaTiSi_3O_9$ Hexagonal

This mineral is unique in more than one way. It is known from only one locality in the entire world—San Benito County in California, from where it gets its name. Until as recently as 1964, when a new mineral (pabstite) was discovered, benitoite was the only mineral known in its particular crystal class, the ditrigonal-dipyramidal class of the hexagonal system. The deep-blue, gem-quality crystals are seen here embedded in fine-grained white natrolite. A few years ago, when the selection of a state mineral for California was being made, benitoite was the closest runner-up to native gold, which is now officially the state mineral.

27. BERYL (variety aquamarine) surrounded by MUSCOVITE

Collection: Krueger
Size: $2\frac{1}{2} \times 2\frac{1}{4}$ in. (6×6 cm.)
Locality: Minas Gerais, Brazil
Beryl: $Be_3Al_2Si_6O_{18}$ Hexagonal
Muscovite: $KAl_2(AlSi_3)O_{10}(OH)_2$ Monoclinic

Six-sided crystals of common beryl as large as a big barrel have been found in some pegmatites, with the barrel-like appearance enhanced because of a series of small, steeply inclined pyramid faces that give to these crystals a slightly tapered outline similar to that of a barrel. These, when capped, as they often are, by a large, flat basal plane, complete the analogy. But it is for the transparent, colored, gem-quality varieties that beryl is best known, under the name of emerald (for green), aquamarine (for blue), morganite (for pink), and heliodor (for yellow).

Aquamarine—as the Latin roots *aqua marina* (water of the sea) imply —is the name given to the variety of beryl that displays a blue color similar to that of the sea. In all its properties, except color, it is identical to emerald, but because aquamarine is less rare than emerald, it does not command as high a price.

The name beryl itself comes from the Greek Berullos, which is believed to have referred to a town in southern India, Belur, near which gemstones were found in ancient times. The opaque, ordinary beryl is often referred to as common beryl (to distinguish it from the many gem varieties). Unfortunately, even "common" beryl is far from common. It is the principal source of the element beryllium (named from the mineral) which, next to lithium, is the lightest of all metals. This feature, coupled with beryllium's considerable structural strength, high melting point, and other characteristics, would make it a "space metal" *par excellence,* were there only more of it. Even despite its high price, beryllium and beryllium alloys are used in many special applications in the aerospace industry. In combination with copper, beryllium makes a notably fatigue-resistant alloy with high electrical conductivity and thus finds important application in the electronics industry. Although some commercial deposits of common beryl occur in South Dakota, Colorado, and a few other places in the United States, the bulk of the industrial supply comes from Brazil.

28. BERYL (variety emerald) in fine-grained PYRITE
matrix

Collection: Land
Size: 3×6 in. (8×15 cm.)
Locality: Muzo, Colombia
Beryl: $Be_3Al_2Si_6O_{18}$ Hexagonal
Pyrite: FeS_2 Isometric

Emerald is the name reserved for the dark-green, transparent, gem-quality variety of beryl. Since the conditions (presence of small amounts of chromium during growth of a beryl crystal) which lead to the formation of the emerald variety are exceedingly rare, it is readily understood why—with its combination of beauty and rarity—emerald has throughout history been one of the most highly valued of all gems, at times exceeding in value even the finest diamonds.

   An area in the province of Muzo, in Colombia (exploited originally by the Incas), is one of two localities in the world that produce the choicest emeralds. The other locality lies in the Ural Mountains of Russia, not far from Sverdlovsk. The name emerald seems to have come, somewhat indirectly, from the old Arabic *zumurrud*, which, however, was applied to any green gemstone, not specifically to emerald.

29. BRAZILIANITE

Collection: McGuinness
Size: 2 × 2 in. (5 × 5 cm.)
Locality: Governador Valadares, Minas Gerais, Brazil
$NaAl_3(PO_4)_2(OH)_4$ Monoclinic

This is one of the more recent additions to the world of minerals, having been discovered and described for the first time in 1945. Transparent crystals are recognized as semiprecious gems. The name reflects the fact that the mineral was first discovered in Brazil.

30. CALCITE

Collection: Halpern
Size: 5 × 7 in. (13 × 18 cm.)
Locality: Joplin, Missouri
$CaCO_3$ Hexagonal

The scalenohedral crystal form that is most characteristic of calcite is here very well displayed. The form takes its name from the fact that each face has the outline of a scalene triangle.

31. CALCITE

Collection: Court
Size: 5×6 in. (13×15 cm.)
Locality: Ouray mining district, Colorado
CaCO₃ Hexagonal

In this specimen, as in the preceding one, the scalenohedral form is dominant.

32. CALCITE

Collection: Halpern
Size: 10×7½ in. (25×19 cm.)
Locality: La Bufa Charcas Mine, San Luis Potosí State, Mexico
CaCO₃ Hexagonal

In this specimen, calcite displays some of its many forms—prisms and rhombohedrons and flat basal pinacoids.

Calcite, in transparent crystals, is known as Iceland spar, and because of its strong double refraction has long been used to produce polarized light, especially in microscopes (see further discussion at plate 36). During World War II, highly flattened crystals (somewhat similar to the disc-shaped crystals on the bottom margin of the photograph here) were discovered in the Santa Rosa Mountains in the desert area of southern California. These found immediate and important application in naval anti-aircraft gunsights and can be credited with warding off attacks on many United States warships.

The name calcite is derived from the Latin *calx* (lime, or limestone).

33. CALCITE with inclusions of HEMATITE

Collection: Smale
Size: 6×6 in. (15×15 cm.)
Locality: Tsumeb, South-West Africa
Calcite: $CaCO_3$ Hexagonal
Hematite: Fe_2O_3 Hexagonal

The red color of the calcite is the result of very finely divided scales of
hematite included within the crystals. Hematite, although a submetallic
mineral, is translucent and brilliant red in thin particles.

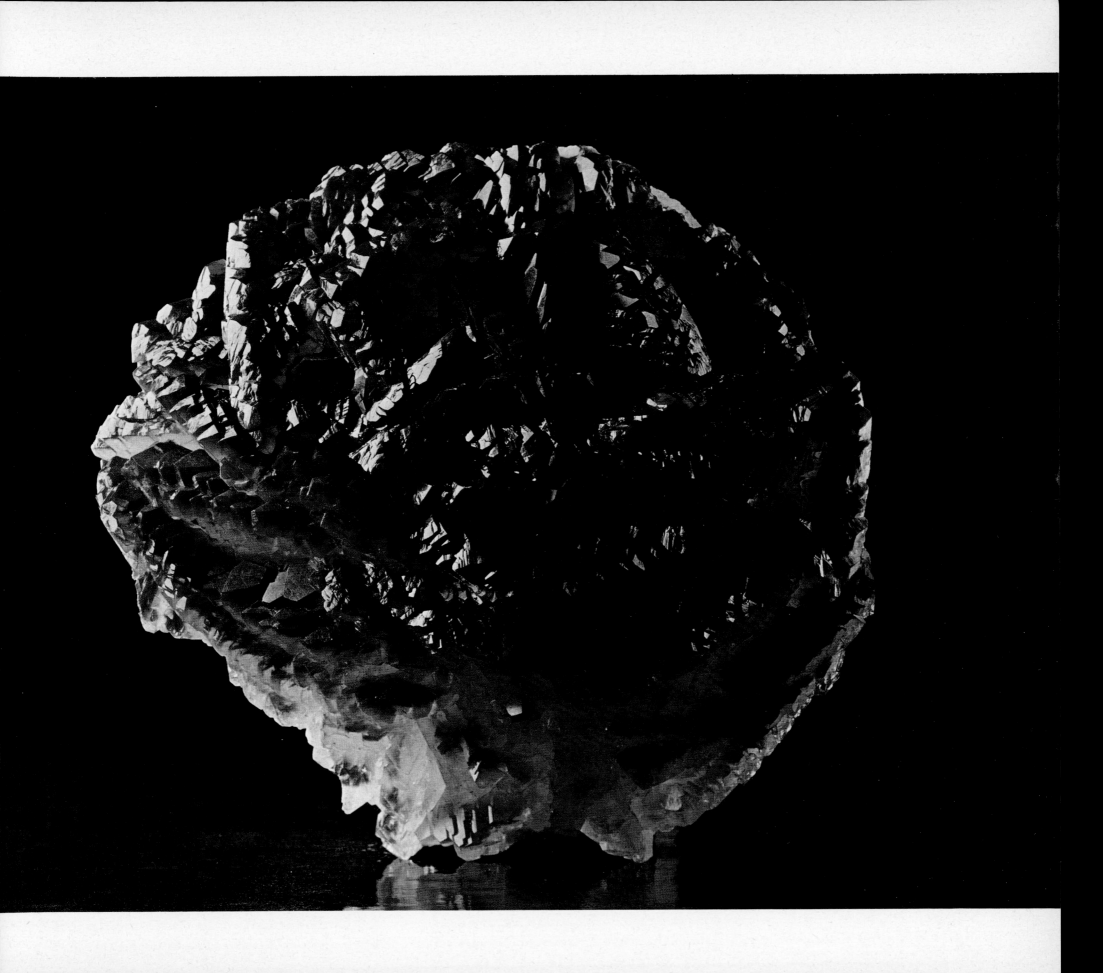

34. CALCITE (replacing an ammonite)

Collection: Court
Size: 3×3 in. $(8 \times 8$ cm.)
Locality: Bavaria, Germany
$CaCO_3$ Hexagonal

Ammonites were abundant members of the mollusk family in the Mesozoic era, some one hundred million years ago. The present-day nautilus may be a descendant of these long-extinct marine animals. Ammonites built their many-chambered shells out of calcium carbonate. Through the millennia, this originally fine-grained material has been recrystallized until now it is virtually a marble—yet still preserving the shape and pattern of the dead animal's "house."

35. CALCITE with LIMONITE

Collection: Court
Size: 5×4 in. (13×10 cm.)
Locality: American Nettie Mine, Ouray mining dis-
              trict, Colorado
Calcite: $CaCO_3$ Hexagonal
Limonite (goethite): $FeO(OH)$ Orthorhombic

Compare this photograph with plate 33. In this case the color is due to a
largely surface coating of other hydrated iron-oxide minerals, which
can conveniently be referred to under the generic term limonite—a
catch-all term for a number of yellow to brown iron oxides, their chief
constituent most commonly being goethite.

36. CALCITE

Collection: Halberstadt
Size: 3 × 9 in. (8 × 23 cm.)
Locality: La Bufa Charcas Mine, San Luis Potosí State, Mexico

$CaCO_3$ Hexagonal

However much one may admire the manifold patterns and colors exhibited by calcite crystals, one should at the same time recognize calcite as a very important rock-forming mineral.

There are in fact very few monomineral rocks in the crust of the earth—that is, rocks composed almost entirely of a single mineral. Olivine in a few places (New Zealand, the northern Cascades of the state of Washington, the southern Appalachians) occurs in masses that may involve a cubic mile or more. Gypsum and halite (common salt) are occasionally found in thin tabular beds extending for many square miles. The calcium-sodium feldspar, labradorite, in a very few places forms bodies of several cubic miles, to which the name anorthosite is given. Quartz sandstones and quartzites of limited extent are known in many places in the geologic section. But calcite, as the predominant and often the only component of limestone and marble, makes up more monomineral rocks than all others put together. Limestones are found widely distributed, both in geologic time and in area. And it is fortunate for man that this is so, for calcite is an extremely important industrial mineral. It is, for example, the essential ingredient in the making of cement, where the action of heat ("calcining") drives carbon dioxide out of the calcite molecule, leaving CaO, a very active chemical, which immediately combines with complex silicates to yield the material that gives Portland cement.

Why is calcite such an abundant and widespread mineral? It is "at home" in all three major rock environments, igneous, sedimentary, and metamorphic (see the Introduction), and it readily precipitates, as an inorganic deposit, from water with a sufficient concentration of calcium and carbonate ions. (The ingredient most commonly responsible for the "hardness" of domestic water supplies is calcium.) But of far more

significance is the fact that calcium carbonate is the preferred skeletal material for the great bulk of invertebrate creatures. The vast coral reefs of the oceans are almost 100 percent calcium carbonate. Clam shells, oyster shells, and many other such structures represent secretion and precipitation, by living matter, of calcium carbonate that has been dissolved in oceans, in lakes, and in river waters. In the course of geologic time, through complex processes, these organic accumulations have been converted to limestone and marbles and now constitute a major earth resource for man. Whenever, therefore, we step onto a cement sidewalk, or into a modern building, or drive on a concrete highway, we should at least figuratively doff our hats in respect to the tiny creatures who in past ages have lived and died by the billions that we may today enjoy such benefits. (Or was that their purpose in life? And what is ours?)

Calcite, in addition to its place in the cement industry, has many other uses, of which only one of the most interesting and specialized will be mentioned—its use in optical instruments. Almost everyone has seen the phenomenon of double refraction. It is best demonstrated by placing a water-clear piece of calcite (known from the locality where it was first discovered and produced as Iceland spar) over a printed word. Viewed through the calcite, the letters will appear double. This optical phenomenon has found practical application in the production of Nicol (or Ahrens) prisms, as efficient devices for turning ordinary light into polarized beams. In recent years polaroid glass has been substituted (because it is cheaper to make and more easily handled and shaped) for many of the applications formerly dependent on Iceland spar. But for many specialized uses in science, Nicol and Ahrens prisms are still preferred.

37. CALCITE with MARCASITE and PYRITE

Collection: Court
Size: 12×7 in. (30×18 cm.)
Locality: Shullsburg, Wisconsin
Calcite: $CaCO_3$ Hexagonal
Marcasite: FeS_2 Orthorhombic
Pyrite: FeS_2 Isometric

The rhombohedron is the dominant form here and gives to the perched calcite crystal its relatively equant shape.

38. CALCITE (variety thinolite)

Collection: Court
Size: 7×8 in. (18×20 cm.)
Locality: Pyramid Lake, Nevada
$CaCO_3$ Hexagonal

This curious pattern is generally regarded as resulting from the replacement of an entire mineral by calcite. In other words, it is a pseudomorph (see the Introduction)—but after what? The problem is still unsolved.

39. CALCITE

Collection: Court
Size: 3 × 4 in. (8 × 10 cm.)
Locality: Arizona
CaCO₃   Hexagonal

This specimen of calcite is best described as vermiform—that is, worm-like. It may be purely of inorganic origin—a sort of stalactitic growth within a cave; it might owe its structure to replacement by calcite of a bundle of grasses or reeds, much as silica replaces logs to yield petrified wood.

40. CASSITERITE

Collection: McGuinness
Size: $2\frac{1}{2} \times 3\frac{1}{2}$ in. (6×9 cm.)
Locality: Acara, Bolivia
SnO_2 Tetragonal

This, the principal ore of tin, is a very distinctive mineral, combining an adamantine luster with a high specific gravity ($G=7$), and having a hardness almost equal to that of quartz ($H=7$). These features, along with a lack of any prominent cleavage and chemical resistance to weathering, ensure that cassiterite will accumulate in placer deposits—as it does most notably along stream channels and even off shore in Malaysia and Indonesia.

The famous deposits in Cornwall were worked by the Phoenicians, perhaps as early as the first millenium B.C.; they were developed more extensively by the Romans during their occupation of southern Britain, and mine production has continued with few interruptions to the present time. The name of the mineral comes directly from the Greek *kassiteros* (tin), which in turns derives from the Elamite language *Kassi-ti-ra,* meaning "from the land of the Kassi" (an Elamite tribe).

41. CELESTITE

Collection: Court
Size: 9 × 9 in. (23 × 23 cm.)
Locality: Matamoros Mine, Coahuila, Mexico
SrSO$_4$ Orthorhombic

42. CELESTITE and CALCITE

Collection: Court
Size: 9½ × 4 in. (24 × 10 cm.)
Locality: Chihuahua, Mexico
Celestite: SrSO$_4$ Orthorhombic
Calcite: CaCO$_3$ Hexagonal

Pale-blue celestite has grown upon and molded around glassy and yellowish calcite.

43. CELESTITE (geode)

Collection: Land
Size: $6\frac{1}{2} \times 7\frac{1}{2}$ in. (17×19 cm.)
Locality: Malagasy Republic
$SrSO_4$ Orthorhombic

Celestite's chemical formula ($SrSO_4$) suggests its close relationship to barite ($BaSO_4$). This close relationship shows also in the similarity of the crystals (like barite, celestite crystallizes in the orthorhombic system) and in the cleavage patterns that both minerals possess. Celestite, although "heavy" by comparison with most nonmetallic minerals (it has a specific gravity of 3.9), does not compare to barite ($G=4.5$). Celestite does not, therefore, find application in drilling muds, nor is it as important an industrial mineral as is barite. It does have one interesting feature: its strontium content; celestite is in fact the principal source of this element.

Strontium alone and in various combinations has the ability to color a flame deep-red. Hence in most red signal flares, strontium is an essential ingredient. The mineral itself very commonly exhibits a delicate pale-blue color (as is well shown in the photograph), and it is from this feature that its name derives. The name celestite comes from the same Latin *caelestis* (sky, or heaven) that has given us the better known English word "celestial" (heavenly).

44. CERUSSITE

Collection: Land
Size: $6 \times 4\frac{1}{4}$ in. (15×11 cm.)
Locality: Arizona
$PbCO_3$ Orthorhombic

As is to be expected, because of the lead content, these white to transparent crystals of cerussite exhibit a truly adamantine luster. The mineral has an index of refraction (see the Introduction) of 2.08, not far below diamond, which has 2.4. Cerussite is a rather minor ore of lead but, as a very distinctive secondary mineral, characteristically forming only at or near the earth's surface, has sometimes been a guide to deposits of primary ore at greater depths.

The name comes from the Latin *cerussa* (waxy) and refers to the somewhat waxy appearance of the crystals, resulting from their adamantine luster.

45. CERUSSITE (reticulated twin aggregate)

Collection: Land
Size: $4\frac{1}{4} \times 6$ in. (11×15 cm.)
Locality: Arizona
$PbCO_3$ Orthorhombic

The interesting pattern exhibited by these crystals results from a type of twinning not uncommon among the orthorhombic carbonates. Technically, the pattern is described as a reticulated aggregate, but more commonly it is referred to as jackstraw structure, because of the resemblance to the patterns in the once popular children's game of jackstraws.

46. CHALCOPYRITE on BARITE

Collection: Halpern
Size: $3\frac{1}{4} \times 3\frac{1}{4}$ in. (8×8 cm.)
Locality: Lauenberg, Germany
Chalcopyrite: $CuFeS_2$ Tetragonal
Barite: $BaSO_4$ Orthorhombic

Tiny tetrahedral forms of chalcopyrite (its characteristic growth habit) are perched on plates of white barite crystals.

47. CHALCOPYRITE and PYRITE

Collection: Court
Size: 4×4 in. (10×10 cm.)
Locality: El Cobre Mine, Zacatecas, Mexico
Chalcopyrite: $CuFeS_2$ Tetragonal
Pyrite: FeS_2 Isometric

In this specimen, cubes of brassy-yellow pyrite are seen nestling in an open pocket of massive chalcopyrite. But compare the color of this chalcopyrite with that of the preceding specimen, in which most of the tiny crystals of chalcopyrite exhibit their familiar and distinctive brassy color—always a somewhat darker brass than that shown by pyrite. What, then, has happened to give this iridescent blue color to this specimen of chalcopyrite? In all probability, nothing more than the development of a thin film—perhaps the result of oxidation—of covellite (CuS) over the surface of the chalcopyrite, just sufficient to replace the yellow of chalcopyrite with the blue of covellite.

Although many minerals—not only sulfides such as chalcocite and bornite, but also arsenides, carbonates, and sulfates—serve as ores of copper, chalcopyrite is today perhaps the most important of the copper ore minerals. This is true because chalcopyrite, in minute amounts, is in many places widely disseminated in massive granitic bodies—the porphyry coppers. The copper content of some of these may be no more than 0.5 percent (only ten pounds of copper in a ton of ore), yet large-scale open-pit mining—as developed, for example, in Arizona's copper belt—makes the mining of such low-grade ore economical.

It is of interest that per-capita consumption of copper, more even than that of iron and steel, is today used as one of the best measures of a nation's industrial development. This application is readily explained when we recall that copper is the major metal used in the electrical industry—in generating, in transmission, and in consumers' appliances; it is also vital in the communications industry and in transportation. And from chalcopyrite we get not only copper but also sulfur which, previously wasted into the atmosphere during smelting of copper-sulfide ores, is now being recovered in increasing amounts both for the prevention of pollution and for its value in the production of sulfuric acid.

The name chalcopyrite comes from the Greek *khalkos* (copper) and from pyrite, which the mineral superficially resembles.

48. CHALCOTRICHITE in CALCITE on native COPPER and LIMONITE

Collection: Halpern
Size: 6×5 in. (15×13 cm.)
Locality: Bisbee, Arizona
Chalcotrichite (cuprite): Cu_2O Isometric
Calcite: $CaCO_3$ Hexagonal
Copper: Cu Isometric
Limonite (goethite): $FeO(OH)$ Orthorhombic

Chalcotrichite, although now known to be identical with cuprite, was at one time believed to be a separate mineral, crystallizing in the orthorhombic system. Typically, chalcotrichite forms in highly elongated, almost hairlike, crystals which, when first studied, were not believed possible in the isometric system. Actually the mineral grows along only one of its "a" axes (see the Introduction, "A Distinctive Crystalline Structure"). In masses this is known as plush copper for obvious reasons. In this specimen, the delicate fibers of chalcotrichite are embedded in calcite, giving the deep-red color to the interior of the calcite crystals. The name comes from two Greek words, *thrix* (hair) and *khalkos* (copper), in allusion to the hairlike growth habit of this variety of cuprite.

49. CHRYSOBERYL

Collection: California Academy of Sciences
Size: $1\frac{3}{4} \times 1\frac{1}{4}$ in. (4×3 cm.)
Locality: Ceylon
$BeAl_2O_4$ Orthorhombic

The name comes from the Greek *khrusos* (gold, or gold-colored) and *berullos* (beryl). Hence, the meaning of the name is a "golden-colored beryl"—though chrysoberyl is a distinct mineral species. The specimen shown here displays both the golden color and an interesting form of twinning—actually a trilling, since three individual orthorhombic crystals interpenetrate in a well-defined geometric pattern to yield the appearance of a hexagonal structure.

50. CHRYSOCOLLA "stalagmites" and drusy
QUARTZ

Collection: Davis

Size: $4\frac{1}{2} \times 3\frac{1}{4}$ in. (11×8 cm.)

Locality: Concepción del Oro, Mexico

Chrysocolla: $Cu_2H_2Si_2O_5(OH)_4$ Monoclinic

Quartz: SiO_2 Hexagonal

The brilliant blue-green of very fine-grained chrysocolla is blended here with a druse (coating of small crystals) of quartz on larger "stalagmitic" shapes.

Chrysocolla (Greek *khrusos,* gold, plus *kolla,* glue) received its name because at one time it was used as a soldering flux for gold.

51. CLEAVELANDITE

Collection: Court
Size: 7 × 8 in. (18 × 20 cm.)
Locality: Sapucaia Mine, Minas Gerais State, Brazil
NaAlSi$_3$O$_8$ Triclinic

Cleavelandite is a variety of albite, the sodium member of the very abundant, very important, and very complex feldspar family of rock-forming minerals. Cleavelandite is the name reserved for albite that occurs in lamellar aggregates, such as are most commonly found in veins and pegmatites.

52. COLUMBITE-TANTALITE with mica

Collection: Court
Size: $5\frac{1}{2} \times 5\frac{1}{2}$ in. (14×14 cm.)
Locality: Berilandia, Conera, Minas Gerais, Brazil
Columbite: $(Fe,Mn)Nb_2O_6$ Orthorhombic
Tantalite: $(Fe,Mn)Ta_2O_6$ Orthorhombic

This is one of the very few mineral names that is almost always hyphenated since virtually all specimens are intermediate in composition between the two end members (see the Introduction): columbite, the niobate; and tantalite, the tantalate.

The mineral is the principal ore of these rare but useful metals. The higher the proportion of tantalum, the more valuable the ore, since tantalum commands a larger market than niobium. It finds particular application in surgical instruments.

Tantalite gets its name from being the principal ore of tantalum, named for the mythical Tantalus, in allusion to the difficulties in making a solution of the mineral in order to get an analysis of its contents. Columbite derives its name from Columbia, once a name for America (after Christopher Columbus); the original specimen was found in Connecticut, from which the element—then known as columbium—was first discovered by Hatchett, in 1802. Subsequently, the name niobium superseded columbium for the element—Niobe being the daughter of Tantalus. But the name of the mineral has continued to be columbite-tantalite, indicating that both elements are almost always present although in varying proportions, from specimens that are almost pure tantalite (which is the more valuable form of the ore) to those that are almost pure niobium.

53. COPPER (native)

Collection: Smale
Size: 3 × 4 in. (8 × 10 cm.)
Locality: Houghton County, Michigan
Cu   Isometric

The tendency of native copper (that is, naturally occurring elemental copper) to form elongated crystals—even though the forms are basically isometric (cubic)—is well shown in this specimen from one of the few mining districts in the world in which the native metal has been the principal ore.

For many years "lake copper" (referring to the Lake Superior area where the mines were located) commanded a premium price on the metal market because nature had produced copper more free from impurities than man was able to, in refining the more widespread and abundant sulfide ores of the "red metal."

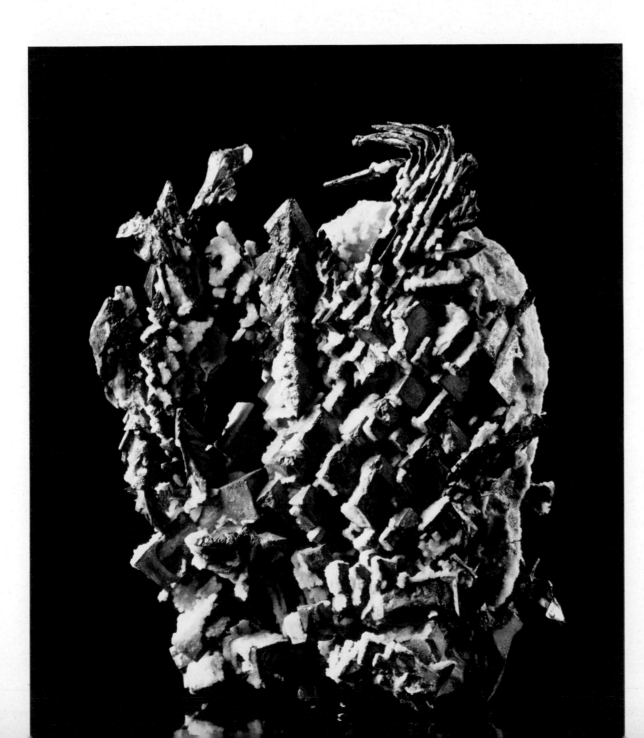

54. COVELLITE with QUARTZ

Collection: McGuinness
Size: 2×3 in. $(5 \times 8$ cm.)
Locality: Leonard Mine, Butte, Montana
Covellite: CuS   Hexagonal
Quartz: SiO_2 Hexagonal

Although not the principal copper ore of this famous mining camp (bornite and enargite have been more important), covellite, with its bright blue color and metallic luster, is one of the most distinctive and easily recognized of the copper minerals. It is named for its discoverer, Nicholas Covelli, an Italian chemist of the nineteenth century.

55. CROCOITE

Collection: Land
Size: $2\frac{1}{2} \times 3\frac{1}{2}$ in. (6×9 cm.)
Locality: Comet Mine, Dundas mining district, Tasmania, Australia

$PbCrO_4$ Monoclinic

Crocoite is a favorite in all mineral displays because of the well-formed and distinctive crystals, the bright orange color (duplicated by almost no other mineral), and the adamantine luster which adds to the brilliance of the color. Although it is a rather rare mineral, found only where weathering of galena (PbS) has brought dissolved lead in contact with some source of chromium, it was in crocoite that a "new metal" (chromium) was first recognized in 1797. The name comes from the Greek word *krokos* for what we know as saffron—the brightly colored orange stigma of a variety of crocus, used today as a dye and as a seasoning.

56. DANBURITE

Collection: McGuinness
Size: 2 × 2 in. (5 × 5 cm.)
Locality: La Bufa Charcas Mine, San Luis Potosí
                State, Mexico

$CaB_2(SiO_4)_2$ Orthorhombic

This mineral is a near relation of topaz, and with its attractive colors, hardness ($H = 7+$), and lack of pronounced cleavage, it has gem qualities in its own right.

The name comes from Danbury, Connecticut, the locality where it was first found.

57. DIOPTASE on limonite-stained QUARTZ

Collection: Land
Size: 3×2 in. (8×3 cm.)
Locality: Tsumeb, South-West Africa
Dioptase: $CuSiO_2(OH)_2$ Hexagonal
Quartz: SiO_2 Hexagonal

Dioptase gets its name from two Greek roots, *dia* and *optazein*, meaning "to see through," because cleavage planes can often be seen within transparent to translucent specimens.

58. DOLOMITE

Collection: Land
Size: $3\frac{1}{4} \times 4\frac{1}{4}$ in. (8×11 cm.)
Locality: Pamplona, Navarra, Spain
$CaMg(CO_3)_2$ Hexagonal

This near relation of calcite ($CaCO_3$) occurs much more frequently in simple rhombs—unlike calcite, which more often exhibits scalenohedral forms (see, for example, plates 30 and 31).

   The name honors the French geologist and alpinist, Déodat de Dolmieu (1750–1801).

59.   DUFTITE on CALCITE

Collection: Krueger
Size: 5×6 in. (13×15 cm.)
Locality: Tsumeb, South-West Africa
Duftite: $PbCuAsO_4(OH)$ Orthorhombic
Calcite: $CaCO_3$ Hexagonal

The mineral was named after G. Duft, a director of the copper mines at
Tsumeb in South-West Africa.

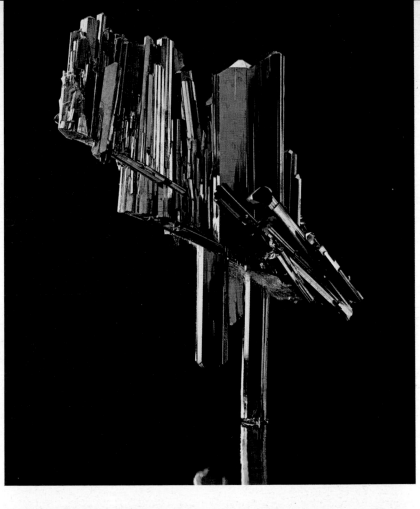

60. EPIDOTE

Collection: Halpern
Size: $2\frac{1}{2} \times 3$ in. $(6 \times 8$ cm.)
Locality: Knappenwald, Salzburg, Austria
$Ca_2(Al,Fe)_3Si_3O_{12}(OH)$ Monoclinic

This mineral was somewhat fancifully named from the Greek words *epi* (additional) and *didonai* (to give—that is, to increase), in recognition of its somewhat unusual growth patterns. Epidote, in transparent and attractively colored specimens, qualifies as a semiprecious gemstone.

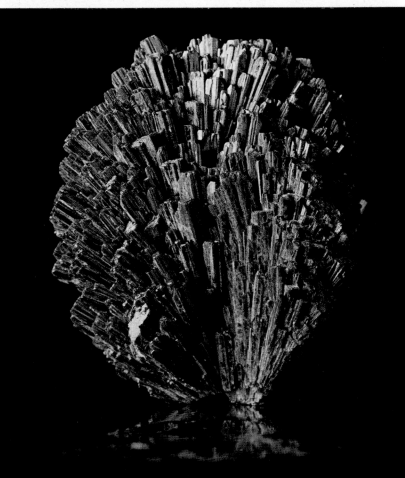

61. EPIDOTE

Collection: Court
Size: $2\frac{1}{2} \times 3$ in. $(6 \times 8$ cm.)
Locality: Mokelumne Hill, California
$Ca_2(Al,Fe)_3Si_3O_{12}(OH)$ Monoclinic

This radiating mass of epidote crystals exhibits a very characteristic growth pattern of this mineral.

63. FERBERITE

Collection: McGuinness
Size: $1\frac{1}{2} \times 2$ in. $(4 \times 5$ cm.)
Locality: Phillips Mine, Nederland, Colorado
$FeWO_4$ Monoclinic

Ferberite is one of the two end members (see the Introduction) of the wolframite series. Huebnerite (see plate 93) is the other. All the intermediate members of this wolframite series are important ores of tungsten.

The name honors the German scientist Rudolph Ferber.

64. FLUORITE on CALCITE

Collection: Davis
Size: $5\frac{1}{2} \times 3\frac{1}{4}$ in. $(14 \times 8$ cm.)
Locality: Le Beix, Puy-de-Dôme, France
Fluorite: CaF_2 Isometric
Calcite: $CaCO_3$ Hexagonal

Fluorite displays here to advantage its characteristic cubic habit, with its transparent crystals, showing in a few places color centers which fade into paler shades throughout the crystal. The calcite, because of its reddish-brown color, obviously contains a small admixture of iron oxide.

65. FLUORITE and CALCITE

Collection: Smale
Size: 6×5 in. (15×13 cm.)
Locality: Province of Asturias, Spain
Fluorite: CaF_2 Isometric
Calcite: $CaCO_3$ Hexagonal

Fluorite is in cubes of unusually dark color; calcite is in its typical white scalenohedrons.

The name fluorite comes from the Latin *fluere* (to flow), because the mineral has a low melting point and is widely used as flux.

66. FORBESITE

Collection: Court
Size: $4 \times 2\frac{1}{2}$ in. (10×6 cm.)
Locality: Ojuela Mine, Mapimí, Durango, Mexico
$H_2(Ni,Co)_2(AsO_4)_2 \cdot 7H_2O$ Monoclinic

Named for David Forbes, an English chemist and geologist (1828–1876).

67. FRANKLINITE in CALCITE

Collection: McGuinness
Size: 4 × 4 in. (10 × 10 cm.)
Locality: Franklin, New Jersey
Franklinite: $(Zn,Mn^{+2},Fe^{+2})(Fe^{+3},Mn^{+3})_2O_4$
Calcite: $CaCO_3$ Hexagonal

A well-formed black octahedron (its edges slightly beveled with the faces of an incipient dodecahedron) is partially embedded in calcite in which the rhombohedral cleavage is well displayed.

68. GALENA

Collection: Court
Size: 3 × 3 in. (8 × 8 cm.)
Locality: Picher, Oklahoma
PbS Isometric

The specimen shows not only the cubic habit but also the characteristic cubic cleavage of galena—the principal ore of lead. The name comes directly from the Latin *galena* (lead ore).

69. GALENA with CALCITE

Collection: Halpern
Size: $5\frac{1}{2} \times 6\frac{1}{2}$ in. (14×16 cm.)
Locality: Joplin, Missouri
Galena: PbS   Isometric
Calcite: $CaCO_3$ Hexagonal

Bullets are no longer made of lead, but when this country was first being settled lead was essential to ammunition. The early discovery of galena in Wisconsin and in the tri-state district (Kansas, Missouri, and Oklahoma) gave enormous encouragement to the western movement in the early history of the United States, for it assured the early settlers of local supplies of lead, whether for shooting game or for protection from Indians.

70. GALENA with MARCASITE and SPHALERITE

Collection: Court
Size: $3\frac{1}{2}\times3\frac{1}{2}$ in. (9×9 cm.)
Locality: Baxter Springs, Kansas
Galena: PbS Isometric
Marcasite: FeS_2 Orthorhombic
Sphalerite: ZnS Isometric

The outer rim of this specimen consists of steel-gray galena (showing well its cubic cleavage); next are brassy-yellow crystals of marcasite; and in the center are ruby-red crystals of sphalerite—a relatively rare variety known to miners as ruby jack.

71. GARNET (variety almandine)

Collection: Smale
Size: 7×12 in. (18×30 cm.)
Locality: Zillerthal Mountains, Austria
$Fe_3Al_2(SiO_4)_3$ Isometric

Garnets are of many varieties, depending on the relative amounts of iron, aluminum, calcium, magnesium, chromium, and manganese that may be present within the consistently isometric structure. Unlike most isometric minerals, garnets rarely develop in the simple, cubic form. Usually they exhibit the more complex dodecahedral (twelve-sided) form shown here. And occasionally garnets may develop as hexoctahedral (forty-eight-sided) forms. The name probably derives from the Latin *granaticus*, a stone so named because its color is similar to the seeds of the pomegranate.

72. GARNET (variety grossular; essonite)

Collection: McGuinness
Size: $1\frac{1}{2} \times 2$ in. $(4 \times 5$ cm.)
Locality: Jeffery Mine, Province of Quebec, Canada
$Ca_3Al_2(SiO_4)_3$ Isometric

The grossular variety of garnet, being relatively free of such coloring agents as iron and manganese which are present in many garnets, is often transparent and—being slightly harder than quartz and having a rather high luster—is prized as a semiprecious gem. The name is derived from old French *grosele* (gooseberry), since the crystals somewhat resemble gooseberries in color, size, and shape. The mineral is in fact frequently called gooseberry stone. Essonite (sometimes hessonite) is the name given to the cinnamon-colored variety.

73. GARNET (variety grossular)

Collection: McGuinness
Size: 2 × 3 in. (5 × 8 cm.)
Locality: Lake Jaco, Chihuahua, Mexico
$Ca_3Al_2(SiO_4)_3$ Isometric

The dodecahedral crystal form here shows to advantage.

74. GARNET (variety andradite)

Collection: Halpern
Size: $2\frac{3}{4} \times 7$ in. (7×18 cm.)
Locality: San Benito County, California
$Ca_3Fe_2(SiO_4)_3$ Isometric

The name andradite was given in honor of the Brazilian geologist José B. de Andrada e Silva (1763–1838).

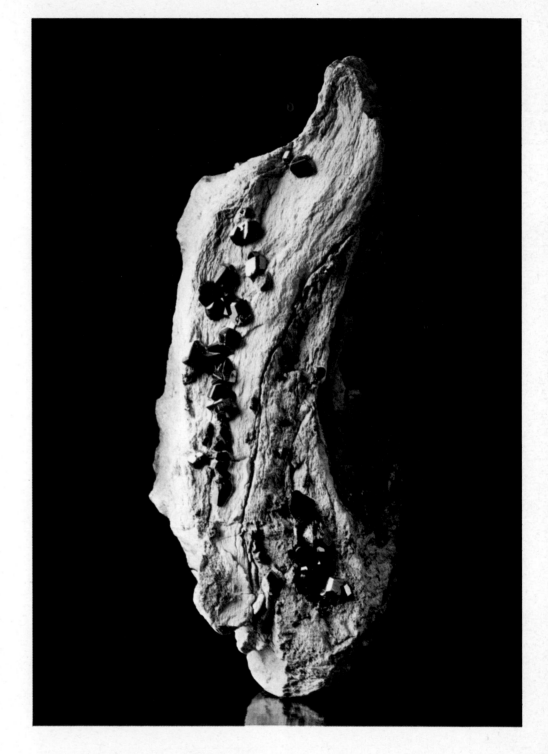

75. GOETHITE (a cast after selenite)

Collection: Court
Size: 9×9 in. (23×23 cm.)
Locality: Naica, Chihuahua, Mexico
Goethite: FeO(OH) Orthorhombic

Here a thin coating of goethite has formed upon a complex of selenite crystals which have subsequently been largely dissolved away—leaving a true cast.

Some iron oxides are black (such as magnetite); some are red to reddish brown (such as hematite); many are yellow and comprise a variety of minerals so fine-grained as to be not readily identifiable. To the latter the all-inclusive name limonite is given. Goethite is usually the principal constituent, and as such is one of the most widespread coloring materials in nature's palette. Wherever we see yellow and yellowish-brown rocks, there probably is goethite.

The name honors the great German poet, Johann Wolfgang von Goethe (1749–1832), who was also a naturalist and at one time was in charge of mining affairs for the Grand Duchy of Weimar.

76. GOETHITE and MALACHITE

Collection: Court
Size: 5×14 in. (13×36 cm.)
Locality: Cole Shaft, Bisbee, Arizona
Goethite: FeO(OH) Orthorhombic
Malachite: $Cu_2(CO_3)(OH)_2$ Monoclinic

The association of goethite and malachite is characteristic in the oxidized zone, that is, the upper, weathered portion of many copper-sulfide ore deposits.

77. GOETHITE on smoky QUARTZ and FELDSPAR

Collection: Court
Size: 4×3 in. (10×8 cm.)
Locality: Crystal Peak, Teller County, Colorado
Goethite: FeO(OH)   Orthorhombic
Quartz: SiO_2 Hexagonal
Feldspar (orthoclase): $KAlSi_3O_8$ Monoclinic

Goethite is in dark, radiating crystal aggregates; feldspar in white crystals; and quartz in glassy, transparent prisms (variety smoky quartz).

78.   GOLD and QUARTZ

Collection: California State Division of Mines and
             Geology, Walter W. Bradley Collection
Size: $2\frac{1}{2} \times 2\frac{1}{2}$ in. (6×6 cm.)
Locality: 200-foot level of the Nigger Hill Mine,
             Jamestown, California

Gold: Au   Isometric
Quartz: SiO_2 Hexagonal

This exceptionally well crystallized gold occurs with white quartz, often called milky quartz, a characteristic gangue mineral of the famous Mother Lode veins.

   Gold and platinum metals are the only ones for which the native element is the principal ore. Silver, copper, mercury, and a few other metals do occur "native" in the crust of the earth, but for none of these is the native metal the most important ore.

79. GYPSUM

Collection: Halpern
Size: 6 × 7 in. (15 × 18 cm.)
Locality: Fremont River Canyon, Wayne County,
              Utah

$CaSO_4 \cdot 2H_2O$ Monoclinic

Gypsum, when heated (calcined), loses part of its water of crystallization. This material, plaster of Paris, when finely ground and mixed
with water, will recombine and quickly harden. It is therefore widely
used as a plaster, in plaster board and other similar applications. The
name comes from the Greek *gypsos* (plaster).

80. GYPSUM (crystal cluster)

Collection: Smale
Size: $13\frac{1}{2}$ × 7 in. (34 × 18 cm.)
Locality: San Quintín, Baja California, Mexico

$CaSO_4 \cdot 2H_2O$ Monoclinic

The somewhat tabular crystals, arranged in a sort of intergrown net-
work, are characteristic of gypsum.

81. GYPSUM rose

Collection: Court
Size: 4×6 in. (10×15 cm.)
Locality: Chihuahua, Mexico
CaSO$_4$·2H$_2$O   Monoclinic

Patterns such as this may develop in gypsum caves and are somewhat akin to the strange growth patterns developed by calcite in limestone caves.

82.   GYPSUM (variety selenite)

Collection: Halpern
Size: $4\frac{3}{4} \times 7\frac{3}{4}$ in. (12×20 cm.)
Locality: Naica, Chihuahua, Mexico
$CaSO_4 \cdot 2H_2O$ Monoclinic

Selenite is the term applied to clear, transparent varieties of gypsum. Contrary to what might be expected, the name does not imply any content of the element selenium. Both names, however, derive from the Greek *selas* (bright), which in turn comes from *selene* (moon).

83. GYPSUM (variety selenite)

Collection: Court
Size: 13 × 10 in. (33 × 25 cm.)
Locality: Cave of Swords, Naica, Chihuahua, Mexico
$CaSO_4 \cdot 2H_2O$ Monoclinic

Study of the cleavage patterns on these well-shaped crystals will reveal that gypsum possesses three different cleavages—unusual in a monoclinic mineral.

84. GYPSUM (variety selenite)

Collection: McGuinness
Size: 3 × 3 in. (8 × 8 cm.)
Locality: Butte, Montana
$CaSO_4 \cdot 2H_2O$ Monoclinic

85. GYPSUM (variety selenite) on CALCITE and
     SPHALERITE

Collection: Court
Size: 8 × 4 in. (20 × 10 cm.)
Locality: Baxter Springs, Kansas
Gypsum: CaSO$_4$·2H$_2$O   Monoclinic
Calcite: CaCO$_3$ Hexagonal
Sphalerite: ZnS   Isometric

In this unusual specimen, tiny bright fibers of selenite are growing from
the surface of brownish-gray calcite scalenohedrons with—in the background—dark, almost submetallic lustered crystals of sphalerite.

86. GYPSUM (variety selenite, fishtail twin)

Collection: Court
Size: 20 × 11 in. (51 × 28 cm.)
Locality: Chihuahua, Mexico
$CaSO_4 \cdot 2H_2O$ Monoclinic

Gypsum not infrequently twins in such a way as to form angles between
crystal terminations not unlike the outline of the tail of some fish—
hence the name fishtail twin.

87. HALITE

Collection: Court
Size: 4×9 in. (10×23 cm.)
Locality: Brawley, California
NaCl   Isometric

The unusual pink color of this ordinarily white or colorless and transparent mineral may be in part due to pink algae incorporated during crystal growth. The name has come from modification of the Greek *als* (sea, or salt), because halite (common salt) is so frequently obtained by evaporation of sea water.

88. HEDENBERGITE

Collection: McGuinness
Size: 3×2 in. (8×5 cm.)
Locality: Langban mining district, Värmland, Sweden
Ca(Fe,Mg)Si₂O₆   Monoclinic

Hedenbergite is a member of the widespread and important pyroxene family of rock-forming minerals.

89. HEMATITE

Collection: Court
Size: 9×7 in. (23×18 cm.)
Locality: Island of Elba, Italy
Fe_2O_3 Hexagonal

Well-crystallized hematite from this world-famous locality is shown here with unusual iridescent colors masking its more usual black color and almost metallic luster.

92.   HEMIMORPHITE and CALCITE

Collection: Court
Size: $5\frac{1}{2} \times 8\frac{1}{2}$ in. (14×22 cm.)
Locality: Mexico
Hemimorphite: $Zn_4Si_2O_7(OH)_2 \cdot 2H_2O$ Orthorhombic
Calcite: $CaCO_3$ Hexagonal

A mass of glassy yellowish crystals of hemimorphite underlie white rhombs of calcite. The name comes from the somewhat curious bipolar forms found on single crystals of the mineral. In this, each end terminates in a different set of faces from the opposite end. Minerals that crystallize in this way (they are uncommon) are said to form hemimorphic crystals because each end of the crystal is only a "half-form." Hemimorphite displays this feature exceptionally well, hence the name.

93. HUEBNERITE

Collection: McGuinness
Size: $2\frac{1}{2} \times 3$ in. (6×8 cm.)
Locality: Adams Mine, San Juan County, Colorado
$MnWO_4$ Monoclinic

Huebnerite is the other end member (see ferberite) of the wolframite series. It was named for Adolph Huebner, a German metallurgist.

94. ILVAITE on CALCITE

Collection: McGuinness
Size: $2\frac{1}{2} \times 3$ in. (6×8 cm.)
Locality: Laxey Tunnel (Golconda claims), South
                 Mountain, Owyhee County, Idaho
Ilvaite: $Ca_2Fe_2^{+2}Fe^{+3}(SiO_4)_2(OH)$ Orthorhombic
Calcite: $CaCO_3$ Hexagonal

The shiny black submetallic crystals are shown to advantage on the background of white calcite. Ilvaite takes its name from the Latin name for the island of Elba, a notable locality for the mineral.

95. INESITE

Collection: McGuinness
Size: $2\frac{1}{2} \times 5\frac{1}{2}$ in. $(6 \times 14$ cm.$)$
Locality: Trinity County, California
$Ca_2Mn_7Si_{10}O_{28}(OH)_2 \cdot 5H_2O$ Triclinic

The pink color on fresh crystal surfaces is characteristic of inesite—as it is of many nonmetallic minerals in which manganese is an important constituent. The name is from the Greek *ines* (flesh fibers), in allusion to the color and structure.

96. LABRADORITE

Collection: Halpern
Size: $4\frac{1}{2} \times 7$ in. (11×18 cm.)
Locality: Labrador, Canada
$(Ca,Na)Al(Al,Si)Si_2O_8$ Triclinic

Labradorite is an important member of the plagioclase series of feldspars—the most abundant minerals in the crust of the earth. Although a number of feldspars occasionally exhibit iridescent surfaces, labradorite is the mineral that most frequently and most intensely shows the play of colors so beautifully displayed in this specimen. The colors are not inherent; they result from reflection and refraction of light from thin layers of the mineral produced by repeated twinnings. Optically, this play of colors is somewhat similar to that in opal, and as with opal, colors will change as light strikes the surface from changing directions.

The name comes from Labrador, where quantities of the mineral have been found.

97. LAUMONTITE

Collection: Court
Size: $10\frac{1}{2} \times 6$ in. (27×15 cm.)
Locality: Pine Creek Mine, Bishop, California
$CaAl_2Si_4O_{12} \cdot 4H_2O$ Monoclinic

Laumontite is a member of the zeolite group of minerals—a group of hydrated alumino-silicates characterized by the fact that, on heating, much of their water of crystallization can be driven off without changing or distorting the crystal structure. Also cation exchange (for example, sodium for calcium) is relatively easily accomplished—hence the use of zeolites as water softeners.

The zeolite group gets its name from the Greek *zein* (to boil). Laumontite is named for the French mineralogist François de Laumont (1747–1834).

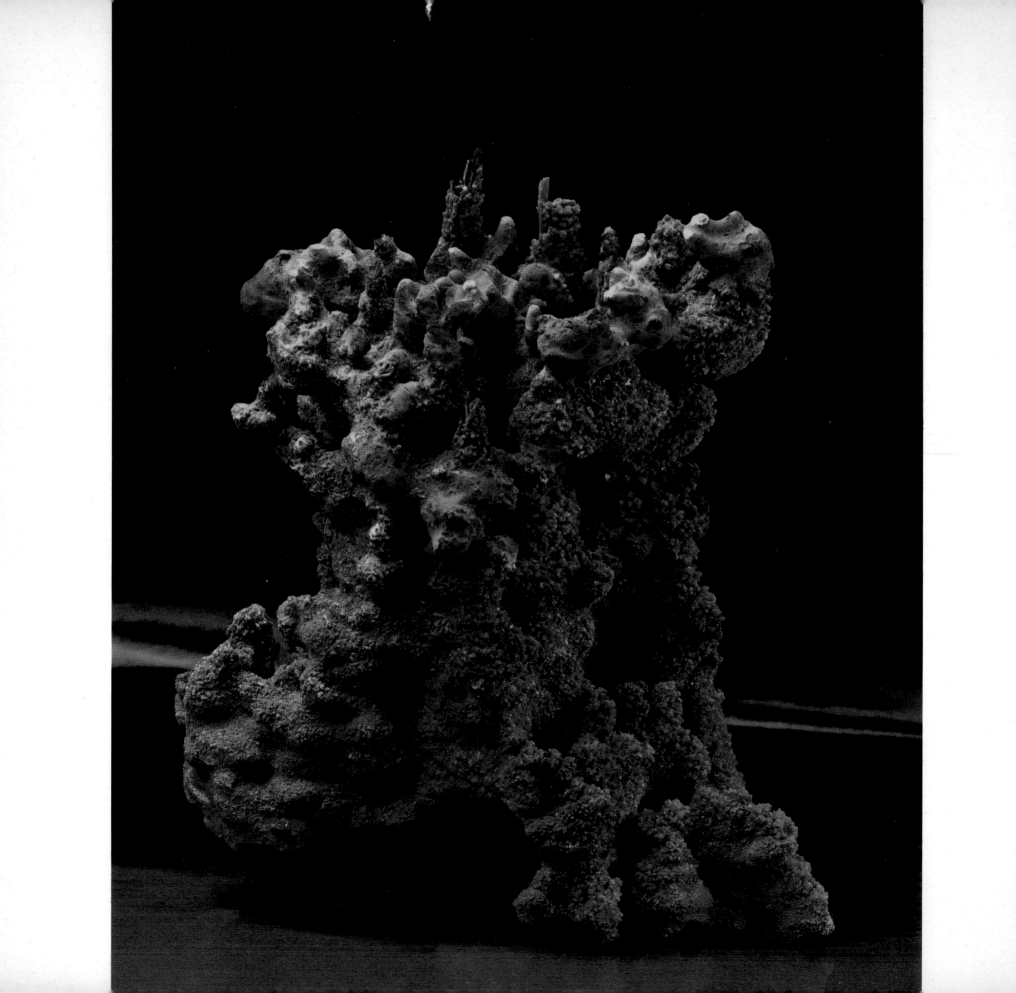

98. LIMONITE (stalactite)

Collection: Halberstadt
Size: 7×9 in. (18×23 cm.)
Locality: Italian Mine, Drytown, California
Limonite (goethite): FeO(OH)   Orthorhombic

Limonite is, as explained under goethite, a mixture of hydrated iron
oxides, of which goethite is usually the principal component. As a
secondary material, it may form pseudomorphs and may take many
strange shapes, including stalactitic forms. The iridescent—often oily-
looking—surface of some boggy ponds is the result of a film of limonite
forming on the surface by evaporation of soluble iron in the water.
Similarly, most rust is merely a film of limonite, developed on a steel or
iron surface because of exposure to moisture in the air. The name comes
from the Greek *leimon* (meadow). Because it is so often found in ponds
in bogs and meadows, the miners' term for limonite is bog iron ore.

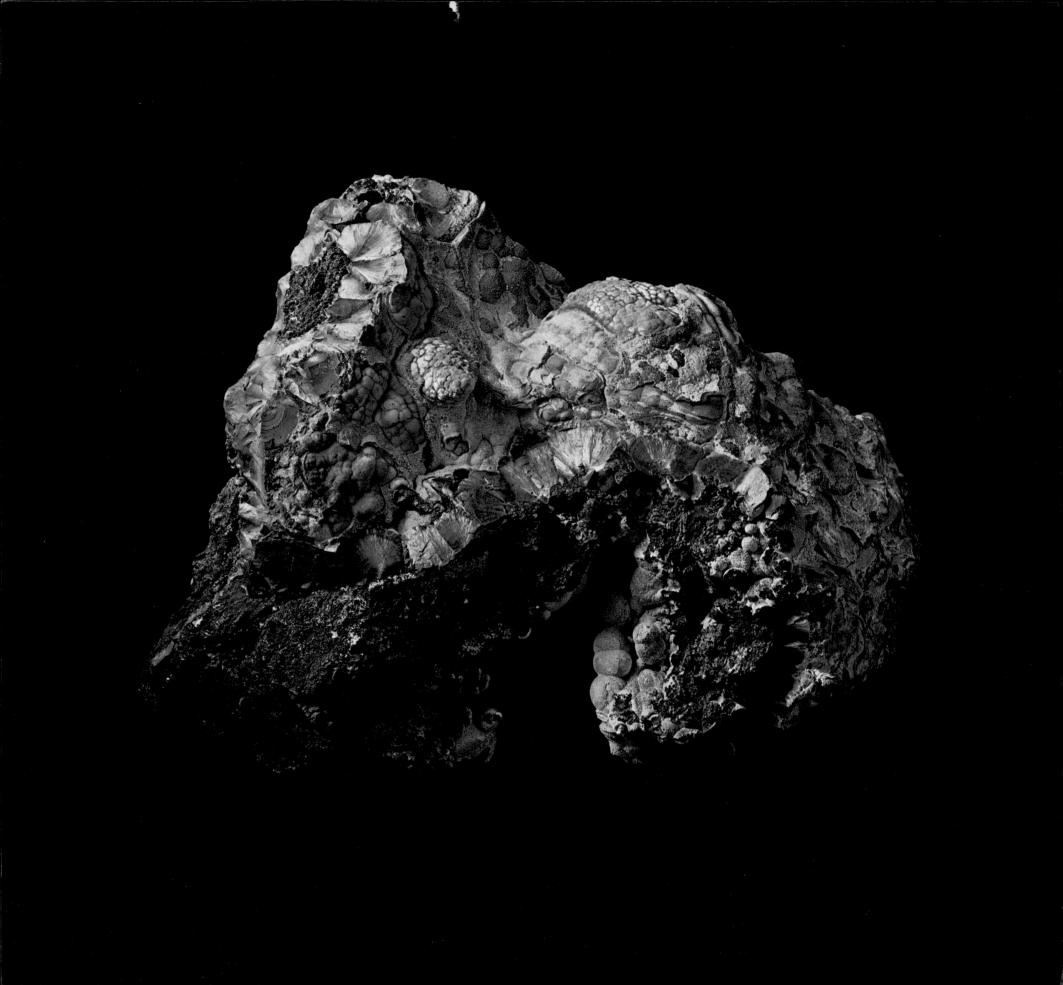

99. MALACHITE

Collection: Court
Size: $9 \times 11\frac{1}{2}$ in. (23×29 cm.)
Locality: Bisbee, Arizona
$Cu_2(CO_3)(OH)_2$ Monoclinic

The radial fibrous structure of the tufts of malachite is here well displayed in some of the broken surfaces. (For more discussion of malachite, see especially plate 18.)

100. MALACHITE (casts after selenite)

Collection: Brown
Size: $6 \times 4\frac{1}{2}$ in. (15×11 cm.)
Locality: Apex Mine, St. George, Utah
$Cu_2(CO_3)(OH)_2$ Monoclinic

The malachite here has so far covered the surface of the original selenite crystals as to give a somewhat rounded appearance to what would otherwise be sharply angled shapes. The malachite is in fact beginning to show its characteristic botryoidal growth habit (that is, resembling a bunch of grapes).

104. MARCASITE ("stalactitic")

Collection: Halpern
Size: $1\frac{1}{2} \times 7$ in. (4 × 18 cm.)
Locality: Ozark Mountains, Missouri
FeS$_2$ Orthorhombic

Stalactites, those remarkable forms that develop as the "dripstone" so characteristic in limestone caves, are most commonly formed by calcite. But when the chemistry of the ground water is right, other minerals may form stalactitic and stalagmitic structures—as is exemplified in this unusual marcasite stalactite from a cave in Missouri.

105. MARCASITE sun

Collection: Davis
Size: 2 × 2 in. (5 × 5 cm.)
Locality: Illinois
FeS_2 Orthorhombic

Marcasite crystals, starting from a central nucleus, have grown outward at such a uniform rate as to form a nearly perfect disc which, with its light brass color, well deserves the name sun.

106.   MESOLITE on THOMSONITE

Collection: Land
Size: 4×7 in. (10×18 cm.)
Locality: Drain, Oregon
Mesolite: $Na_2Ca_2(Al_6Si_9)O_{30} \cdot 8H_2O$ Monoclinic
Thomsonite: $NaCa_2(Al_5Si_5)O_{20} \cdot 6H_2O$ Orthorhombic

The mesolite crystals show here, as in plate 107, their characteristic elongated growth habit. In this specimen their diameter is scarcely more than a hair's width!
    Thomsonite is another member of the zeolite group.

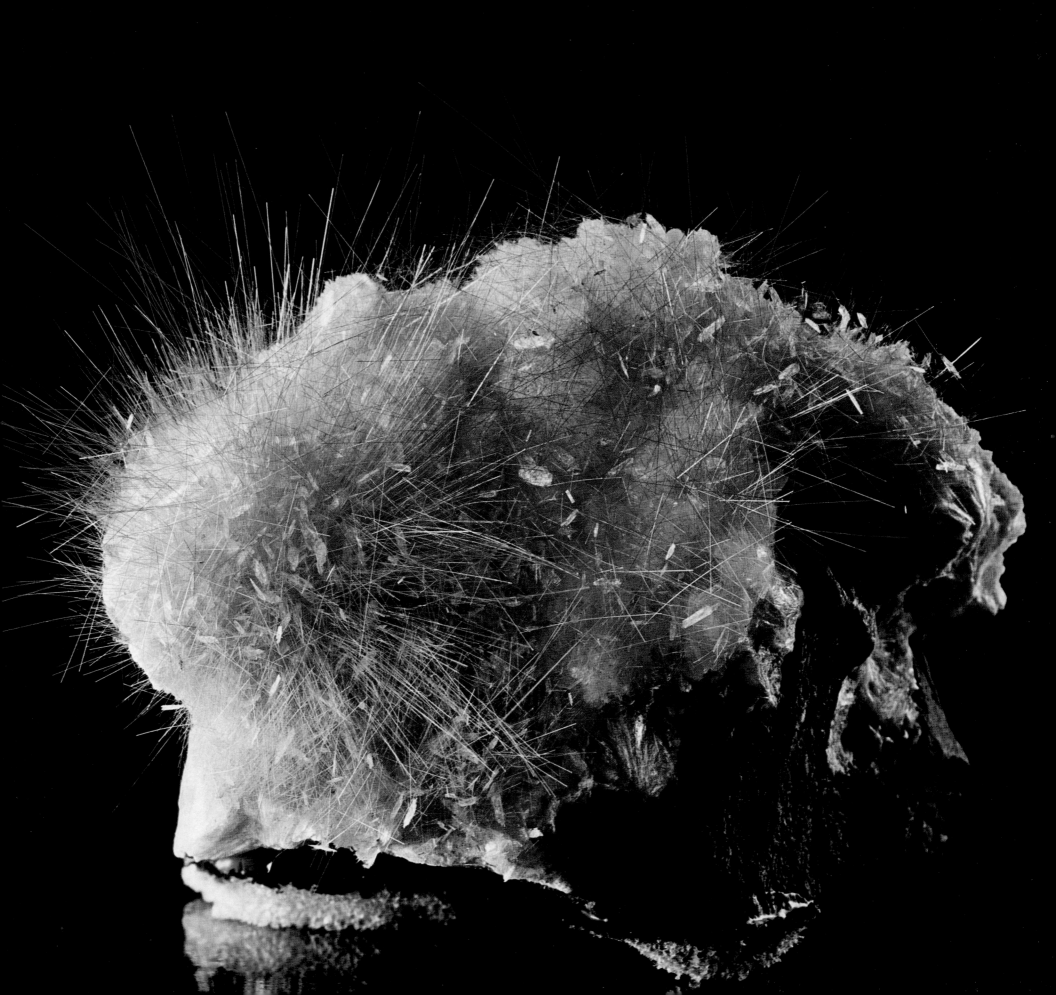

107.   MESOLITE on STILBITE

Collection: Land
Size: 7×4 in. (18×10 cm.)
Locality: Bucoda, Washington
Mesolite: $Na_2Ca_2(Al_6Si_9)O_{30} \cdot 8H_2O$ Monoclinic
Stilbite: $NaCa_2(Al_5Si_{13})O_{36} \cdot 14H_2O$ Monoclinic

Mesolite, like laumontite, is a member of the zeolite group. It takes its name from the Greek word *mesos* (middle) because mesolite, when first discovered and analyzed, seemed to fall between scolecite, a calcium zeolite, and natrolite, a sodium zeolite. Stilbite comes from the Greek *stilbos* (shimmering), in reference to the shiny, white, translucent crystals.

108.   MICROCLINE (variety amazonite) and smoky
       QUARTZ

Collection: Land
Size: $3\frac{1}{4} \times 4\frac{1}{2}$ in. (8×11 cm.)
Locality: Pike's Peak area, Colorado
Microcline: $KAlSi_3O_8$ Triclinic
Quartz: SiO_2 Hexagonal

Microcline is a member of the feldspar family, identical in its composition with the better known orthoclase but differing from it in being
triclinic (orthoclase is monoclinic). Moreover, microcline occasionally
—perhaps as the result of some ferrous iron in its structure—develops a
distinctive green color which makes it prized as the semiprecious stone
called amazonite. The name comes from the Amazon River in Brazil,
where green stones were reported by early explorers—though these
were probably not amazonite as we know it today.

109.   MIMETITE

Collection: Land
Size: $3\frac{3}{4} \times 4\frac{3}{4}$ in. (10×12 cm.)
Locality: San Pedro Corallitos, Chihuahua, Mexico
$Pb_5(AsO_4)_3Cl$ Hexagonal

The name mimetite comes from the Greek *mimes* (imitator), because
the crystal shapes of mimetite imitate those of a more abundant and
closely related species, pyromorphite.

110.   MIMETITE and PLUMBOGUMMITE

Collection: McGuinness
Size: $1\frac{1}{2} \times 2\frac{1}{2}$ in. (4×6 cm.)
Locality: Alston Moor, Cumberland, England
Mimetite: $Pb_5(AsO_4)_3Cl$ Hexagonal
Plumbogummite: $PbAl_3(PO_4)_2(OH)_3 \cdot H_2O$ Hexagonal

The bright yellow of mimetite contrasts with the deep blue of plumbogummite, which derives its name from the Latin *plumbum* (lead) and
gummi (gum) because in some occurrences the mineral resembles drops
of gum.

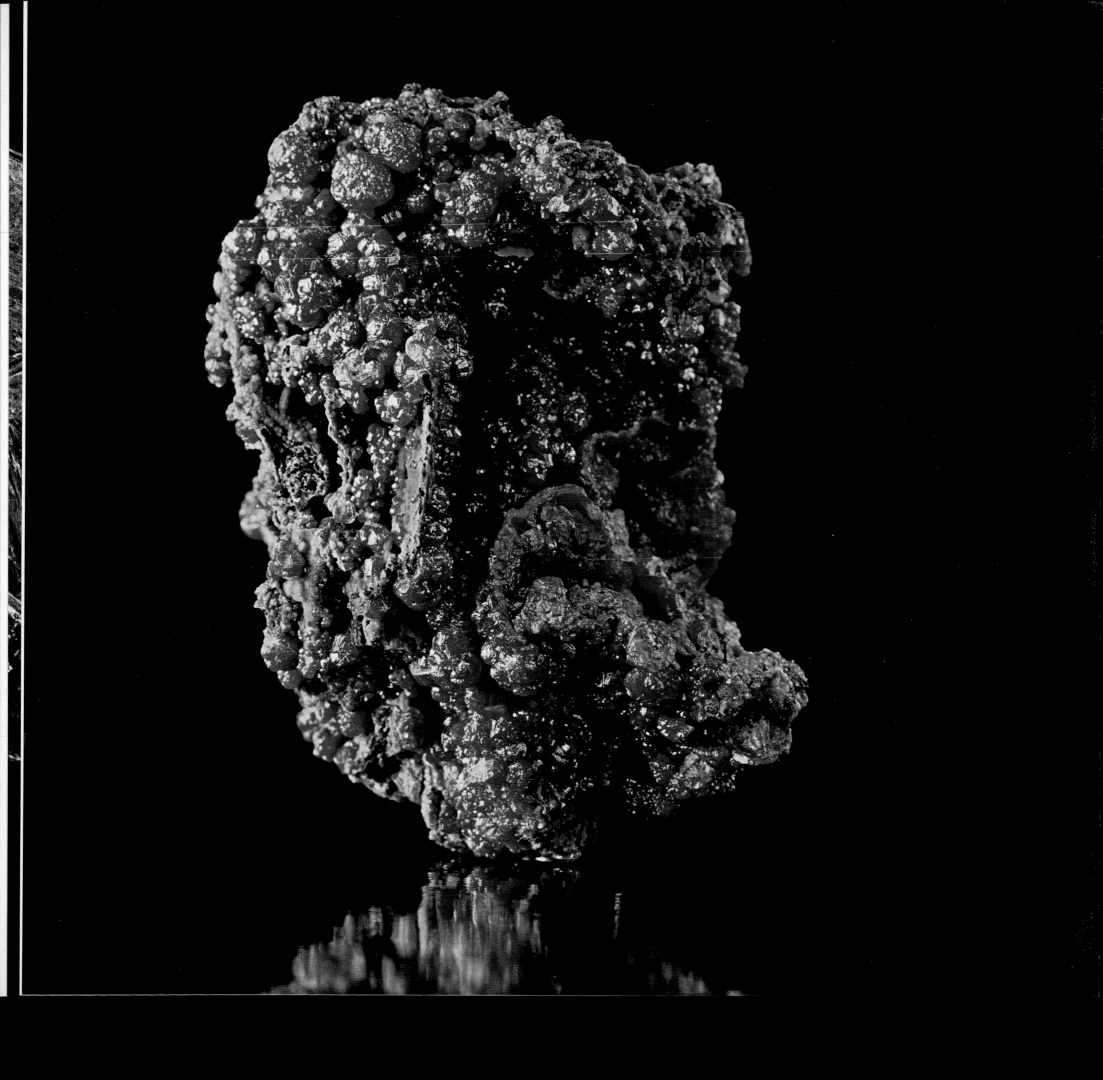

117. PECTOLITE

Collection: Court
Size: 4×4 in. (10×10 cm.)
Locality: Paterson, New Jersey
$NaCa_2Si_3O_8(OH)$ Triclinic

118.   PREHNITE on PECTOLITE

Collection: McGuinness
Size: 5×3 in. (13×8 cm.)
Locality: Prospect Park, New Jersey
Prehnite: $Ca_2Al_2Si_3O_{10}(OH)_2$ Orthorhombic
Pectolite: $NaCa_2Si_3O_8(OH)$ Triclinic

Aggregates of pale-green prehnite are here growing on and around
fibrous masses of white pectolite. Prehnite was named for the Dutch
Colonel von Prehn who, in 1774, first brought the mineral, from a
locality near the Cape of Good Hope, to the attention of European
mineralogists.

119. PYRITE

Collection: Halpern
Size: 2 × 2 in. (5 × 5 cm.)
Locality: Dallas, Texas
FeS₂ Isometric

Specimens like this were found in the course of construction excavation for the Dallas, Texas, airport. First, a concretionary mass of fine-grained pyrite had formed within a bed of shale; then, with some slight change in growing conditions, large single cubes were able to form on the outer surface of the spherical concretion to produce this striking pattern.

120.   PYRITE (octahedron)

Collection: Court
Size: 3×5 in. (8×13 cm.)
Locality: Peru
FeS_2 Isometric

Pyrite here displays another of its characteristic crystal forms, the eight-sided octahedron, in which each face is an equilateral triangle.

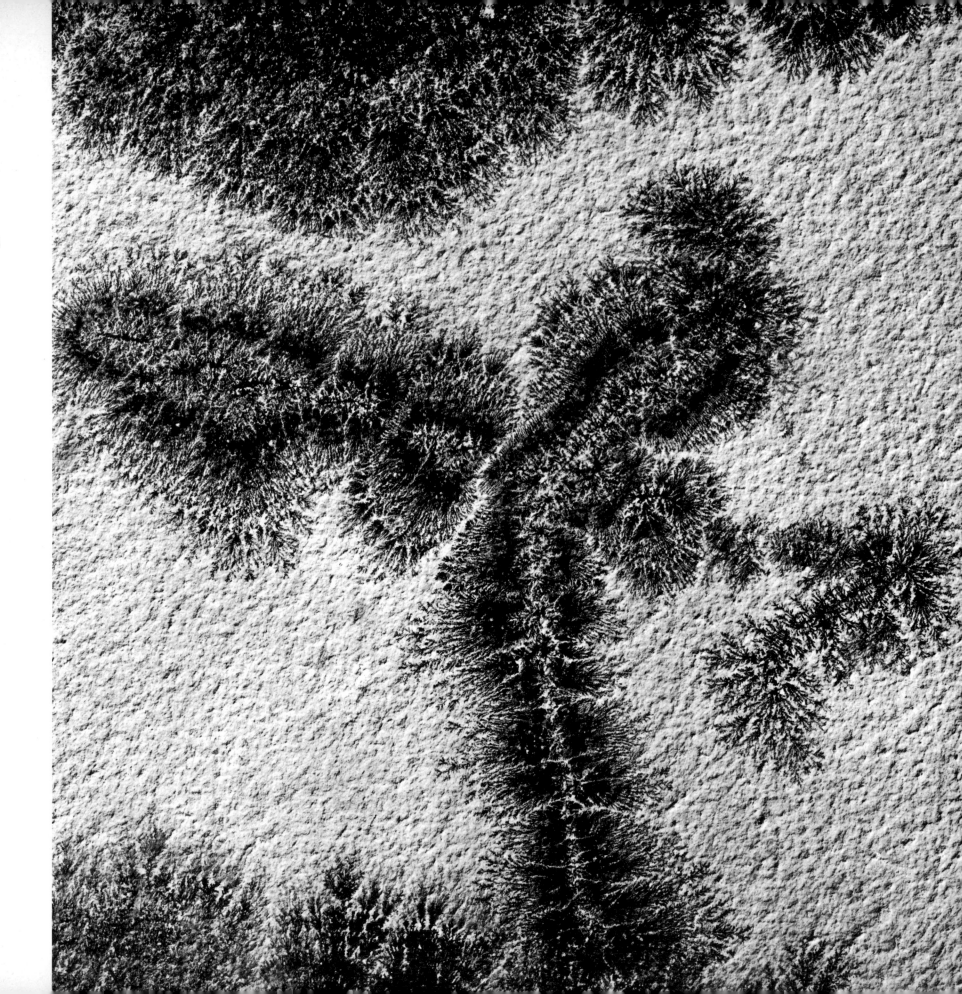

216

124. PYROLUSITE (dendrites)

Collection: Smale
Size: 12 × 12 in. (30 × 30 cm.)
Locality: Solnhofen, Bavaria, Germany
MnO_2 Tetragonal

Pyrolusite here exhibits its tendency to form dendritic growth patterns (see the Introduction).

125. PYROMORPHITE

Collection: McGuinness
Size: 1 × 3 in. (3 × 8 cm.)
Locality: Germany
$Pb_5(PO_4)_3Cl$ Hexagonal

This mineral, along with mimetite (see plates 109 and 110), forms one of the more unusual solid-solution series in the mineral world—that is, a series in which the chemical composition can change without serious distortion of the crystal form. Thus sodium and calcium may interchange in feldspars, iron and magnesium in olivines, and so on. But it is rare that cation complexes can substitute one for another, as happens here, where—if phosphate is replaced by more than 50 percent of arsenate—the mineral becomes mimetite.

Pyromorphite gets its name from the Greek *pyr* (fire) and *morph* (form) because pyromorphite when fused (it melts easily because of its lead content) forms rounded globules which, on cooling, take on crystalline form.

126. PYRRHOTITE with CALCITE and SPHALERITE

Collection: McGuinness
Size: 2 × 2 in. (5 × 5 cm.)
Locality: Potosí Mine, Aquiles Serdán (formerly known as Santa Eulalia), Mexico
Pyrrhotite: $Fe_{1-x}S$ Hexagonal
Calcite: $CaCO_3$ Hexagonal
Sphalerite: ZnS Isometric

The curious formula given above for pyrrhotite expresses the fact that in most analyzed specimens there is never quite enough iron to make a one-to-one relationship with sulfur. Interestingly, in one variety of pyrrhotite, troilite, which is known principally from its occurrence in meteorites, the ideal formula of FeS is reached. Pyrrhotite gets its name from the Greek *pyrrhotes* (redness), in allusion to the reddish bronze color of the mineral on a freshly broken surface.

127. PYRRHOTITE and QUARTZ on SIDERITE

Collection: Smale
Size: 2×4 in. (5×10 cm.)
Locality: Moro Velho Mine, Minas Gerais, Brazil
Pyrrhotite: $Fe_{1-x}S$ Hexagonal
Quartz: SiO_2 Hexagonal
Siderite: $FeCO_3$ Hexagonal

The bronze metallic crystals of pyrrhotite, along with glassy quartz, are
perched on brown, translucent rhombs of siderite.

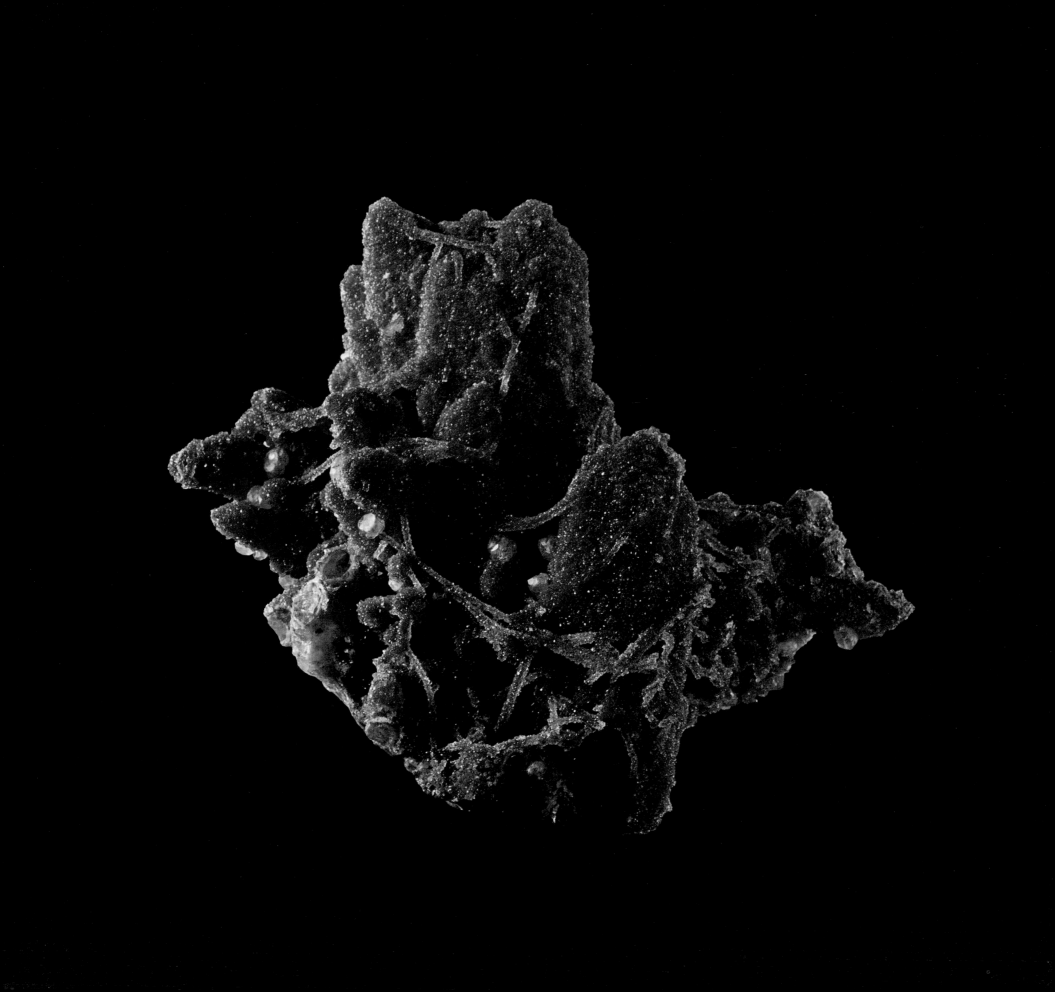

128. QUARTZ

Collection: Court
Size: $5\frac{1}{2} \times 4$ in. (14×10 cm.)
Locality: Zacatecas, Mexico
SiO_2 Hexagonal

The unusual brilliant red color here results from the inclusion of tiny scales of hematite (see, for example, plates 89–91) within the quartz.

Quartz is the mineral more often found throughout the earth's crust in recognizable crystals than any other. In addition, of all the minerals in the world, it is the one that has served man the longest, and with a wider range of uses. In the form of flint, quartz was probably the mineral used by primitive man over a million years ago for some of his first crude scrapers and grinders. Today, delicate quartz-crystal "wafers" govern the frequencies in radio and electronic equipment and make it possible for men to fly to the moon and back. Throughout all these years, in one or another of its varied manifestations, quartz has not only served the artisan and the engineer but has also intrigued the scientist and attracted the artist. It is one of the very few minerals that has had whole books devoted to it (see the recommended reading list).

What are some of the properties that make quartz distinctive? Perhaps its most obvious feature, when it is well crystallized (and it often is), is its hexagonal form; moreover, it is the most common mineral in the hexagonal crystal system. But the seemingly simple hexagonal prism, which is most readily recognized, is really not so simple! Here we cannot go into details, but the fact is that quartz belongs to a very special class of this system—the trapezoidal. Not only that, quartz may come in either left-handed or right-handed forms—indicated by small crystal faces that often do not appear on some crystals. Such crystals are so named because they have the ability to rotate the plane of a polarized light beam either to the left or to the right, depending on whether the crystal is right- or left-handed. It is well known that sugar also has left or right forms—referred to as levulose and dextrose, respectively; thus quartz crystals, since this property was discovered in them, have been used to measure the strength of sugar solutions.

But long before any of the industrial developments mentioned took place, quartz crystals had been studied by a noted Danish scientist, Nicolas Steno, who in 1669 announced the "law of the constancy of interfacial angles." He had observed that no matter how large or small a crystal of quartz, whether it was long and narrow or short and thick, the angles between the prism faces remained constant, always 60 degrees. From this important observation came the beginnings of the science of crystallography, which has led into modern developments of X rays and all that they have revealed about the structure of matter—including man.

There are other interesting features in connection with the crystallography of quartz. Because of its unique structure it possesses piezoelectric properties, which means that, if quartz of the right shape is subjected to pressure in the right direction, it will generate, momentarily, a small electric current. This can be adjusted very precisely by proper shaping and design. Hence quartz oscillators are used to govern the frequency of all kinds of electronic gear. There is an opposite effect, too: a small electric current, applied to a quartz crystal, will generate a pressure wave. This feature was used extensively in the first acoustic-ranging devices in submarine detection. In addition, let us not forget that quartz is the principal ingredient in such mundane but important things as sandpaper, whetstone, silica brick (used in steel-making), and countless other devices dependent on two of its qualities: its hardness (quartz easily scratches glass and hence has been used as fake diamonds) and its lack of cleavage, so that when broken it presents sharp edges rather than smooth faces.

Quartz has gem value as well, in many, different-colored varieties—amethyst, chrysoprase, cairngorm, rose, citrine, bloodstone, and others—which are widely distributed on the earth. Most of the best amethysts, for example, come from Brazil and the Ural Mountains in Russia; chrysoprase is found in Australia and in California; cairngorms (as the name implies) occur typically in the Cairngorm Mountains of Scotland; rose quartz is found in South Dakota, New England, and elsewhere.

Considering quartz's importance, wide distribution, and distinctive features, it is a bit surprising and certainly disappointing that neither mineralogists nor etymologists have yet been able to provide a definitive explanation of how quartz came to be known by this name. There is little doubt that the material referred to as *krystallos* (ice) by Greek and Roman writers was what today we call rock crystal—that is, clear, well-crystallized quartz. (The ancient Greeks were familiar with quartz crystals and knew that the specimens they saw had been found mostly to the north, especially in the Alps. They thought of this as cold country, and—partly because of the colorless appearance of these crystals—believed them to be ice that had been frozen so hard it would never melt!) But that the finer-grained forms and variously colored varieties of SiO_2 were in any way related to rock crystal, much less that they were essentially identical to it, did not become known until many centuries later. In the meantime, the name was beginning to appear in German writings of the sixteenth century (either as *quarz* or *querz*) in reference to the gangue (non-ore material) of mineral veins. One suggestion is that the name is an abbreviation of an Old Saxon mining term, *querklufterz* (cross-vein ore). Another is that the name derives from Old Slavic *tvruda* (hard), which in the course of transliteration through Czech and Polish became *quarz* in German. In any event, the name was not then applied to rock crystal but to any hard, dense, siliceous material found in ore veins. In the Anglicized form of "quartz," the name does not appear in English writings until the seventeenth century, and it did not come into general use until well into the eighteenth century. Through the Greek word *krystallos*, the name of the science of crystallography was derived.

129. QUARTZ (variety amethyst) and EPIDOTE

Collection: Court
Size: $7 \times 3\frac{1}{2}$ in. (18×9 cm.)
Locality: Veracruz, Mexico
Quartz: SiO_2 Hexagonal
Epidote: $Ca_2(Al,Fe)_3Si_3O_{12}(OH)$ Monoclinic

Amethyst is the name given to the deep-blue to violet, gem-quality quartz. Amethyst is therefore identical with quartz in all of its physical and chemical properties. The color is at least in some instances the result of the incorporation of thin films of iron oxide on the faces of the crystals as they were growing. The color may sometimes be destroyed by heating or may be changed to the yellow of citrine. The name derives from the Greek *amethustos* (anti-intoxicant) because the wearing of an amethyst was at one time believed to cure—if not also to prevent—hangovers. Whether true or not, this belief, along with the natural beauty of the mineral, has made amethyst a prized gemstone from antiquity right down to the present day.

130. QUARTZ (with parallel inclusions)

Collection: Court
Size: $4\frac{1}{2} \times 4\frac{1}{2}$ in. (11×11 cm.)
Locality: Minas Gerais, Brazil
SiO_2 Hexagonal

The layer-like pattern is the result of weathering out of small amounts of clay minerals that were not replaced during the initial growth of the quartz.

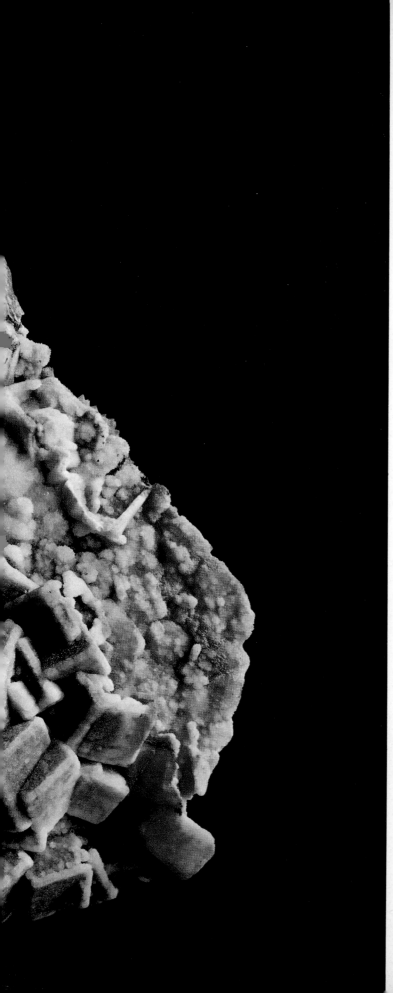

131. QUARTZ (cast after fluorite)

Collection: Court
Size: $17\frac{1}{2} \times 13$ in. (44×33 cm.)
Locality: Idarado Mine, San Juan mining district, San
                    Juan County, Colorado
SiO_2 Hexagonal

This is quartz, but it is not cubic! The cubic appearance is simply the result of a quartz druse having been deposited on a cluster of fluorite cubes, with subsequent removal, by natural solutions, of virtually all of the fluorite.

132. QUARTZ

Collection: Court
Size: 4×4 in. (10×10 cm.)
Locality: Veracruz, Mexico
SiO_2 Hexagonal

This spectacular grouping of quartz crystals indicates why this mineral
is often referred to as rock crystal.

133. QUARTZ (variety chalcedony)

Collection: Court
Size: $6\frac{1}{2} \times 5$ in. $(17 \times 13$ cm.)
Locality: Crook County, Oregon
SiO$_2$ Hexagonal

This specimen of chalcedony shows a vermiform (wormlike) growth coarser than that of the calcite specimen shown at plate 39. The mineral developed in a bubble of gas frozen in rhyolitic lava.

134. QUARTZ (variety chalcedony rose)

Collection: Court
Size: 2×4 in. $(5 \times 10$ cm.)
Locality: Minas Gerais, Brazil
SiO$_2$ Hexagonal

The finely fibrous texture of the chalcedony is seen here to advantage. The whole bears some resemblance to a rose—hence the name.

136. QUARTZ (crystals showing parallel growth)

Collection: Smale
Size: $6 \times 5\frac{1}{2}$ in. $(15 \times 14$ cm.$)$
Locality: Goiás, Brazil
SiO_2 Hexagonal

138. QUARTZ (variety flint)

Collection: Land
Size: 7×8 in. (18×20 cm.)
Locality: Dover, England
SiO_2 Hexagonal

Flint most commonly forms as concretions (see the Introduction) within limestone or shale—hence the strange shapes that it frequently exhibits. Flint is not really harder than other varieties of quartz, although it may seem so because, being extremely fine-grained, it is also extremely tough. The fine-grained, almost amorphous texture permits the shaping of flint, whether into rounded mortars and pestles for grinding, into sharp edges for cutting, or into points for spears and arrowheads. Flint was therefore one of the most prized of minerals during the Stone Age. The name appears to have come directly from the Anglo-Saxon *flint*.

139. QUARTZ (variety rock crystal)

Collection: Court
Size: $5\frac{1}{2} \times 5$ in. (14×13 cm.)
Locality: Herkimer County, New York
SiO_2 Hexagonal

These well-shaped crystals have formed within small cavities in limestone. Because of their sparkling transparency they were at one time sold to unsuspecting customers as diamonds. But they are collector's items in their own right—as excellent examples of well-crystallized quartz. Because of their past history and their occurrence in Herkimer County, they have become widely known as Herkimer diamonds.

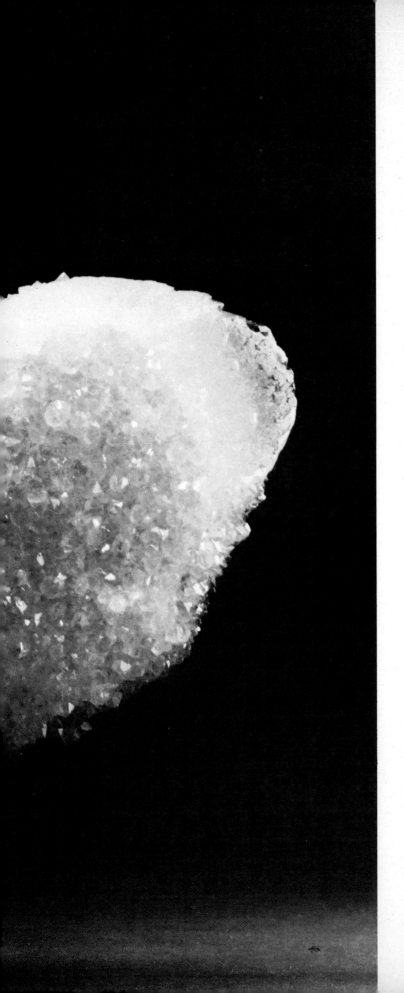

140. QUARTZ (Japanese twin)

Collection: Davis
Size: $7\frac{1}{2} \times 7$ in. (19×18 cm.)
Locality: The Holland Mines, Washington Camp,
             Arizona

SiO_2 Hexagonal

Quartz may twin in any of several geometric patterns to which the name twinning laws is given, and which are differentiated by different names for each law. In the Japanese law—so-called because twin crystals of this type were first found in Japan—the twin plane (the surface of junction) is parallel to the face of the dipyramid.

141. QUARTZ

Collection: Court
Size: 5 × 8 in. (13 × 20 cm.)
Locality: Bagdad, Arizona
SiO_2 Hexagonal

142. QUARTZ (variety rose quartz) on smoky quartz

Collection: California Academy of Sciences
Size: $11\frac{1}{2} \times 10$ in. $(29 \times 25$ cm.)
Locality: Governador Valadares, Minas Gerais, Brazil
SiO_2 Hexagonal

The specimen is unusual in that rose quartz—ordinarily found in massive form—occurs here in well-faceted crystals and also in combination with smoky quartz. The color of rose quartz is believed to result from the scattering of incident light by tiny, oriented crystals of rutile (TiO_2).

The cause of the smoky color in quartz has been investigated repeatedly, with many hypotheses being advanced and later discarded as more accurate structural and chemical data have become available. Although the matter cannot be regarded as settled, there is good reason to believe that a small content of aluminum is a prerequisite; nevertheless, some colorless quartz contains as much and more aluminum as does smoky quartz. Thus it is not the amount of aluminum incorporated in a quartz crystal during growth that confers the smoky color but rather the positioning of the aluminum atoms within the quartz structure during its growth. The smoky color can sometimes be developed by exposing colorless quartz to X rays, provided that some aluminum is present and in the right places within the crystal.

One of the best-known localities for smoky quartz is the Cairngorm Mountains of Scotland. Smoky quartz of the particular shade of smokiness characteristic of this area is used as a semiprecious gemstone, and the name cairngorm has become almost synonymous with this variety of smoky quartz. It has in fact almost become the national stone of Scotland, where it is widely used in brooches and other settings.

143. QUARTZ (smoky quartz) with etched FELDSPAR

Collection: Court
Size: $10 \times 12\frac{1}{2}$ in. (25×32 cm.)
Locality: Shaver Lake, Fresno County, California
SiO_2 Hexagonal

The horizontal lines that give an interesting pattern to the otherwise smooth vertical prism faces in this crystal result from small oscillations during the growth of the crystal: oscillations between the upward growth of the prism face and the slantward growth of the rhombohedron face above. The intersection, because of the perfection of the crystal, is a tiny edge which appears as a horizontal line, repeated as many times as there were minor oscillations during growth.

144. QUARTZ (sand casts after HALITE)

Collection: Krueger
Size: $4 \times 4\frac{1}{2}$ in. (10×11 cm.)
Locality: Brawley, California
SiO_2 Hexagonal

In this case the cubic pattern of the quartz druse results, not from a cast after fluorite but from a cast after halite (NaCl), which also is a cubic mineral.

145. REALGAR and CALCITE

Collection: Court
Size: $5\frac{1}{2} \times 3$ in. (14×8 cm.)
Locality: Getchell Mine, Humboldt County, Nevada
Realgar: AsS Monoclinic
Calcite: $CaCO_3$ Hexagonal

Ruby-red realgar crystals contrast here with white calcite, locally darkened by traces of sulfides. Realgar takes its name from the Arabic *rajh al-ghar* (powder of the cave, or mine), in reference to the fact that on long exposure to light the mineral may disintegrate to a red-yellow powder.

146. RHODOCHROSITE

Collection: Court
Size: $4\frac{1}{2} \times 6$ in. (11×15 cm.)
Locality: Silverton Tunnel, Ouray, Colorado
$MnCO_3$ Hexagonal

The angles between the rhombohedral cleavages of this mineral, here so well displayed, are not far from 90 degrees, thus giving an almost cubic appearance. The name comes from the Greek *rhodon* (rose) and *khrosis* (color).

147. RHODONITE with CALCITE

Collection: McGuinness
Size: $2\frac{1}{2}\times 3$ in. (6×8 cm.)
Locality: Franklin Furnace, New Jersey
Rhodonite: $MnSiO_3$ Triclinic
Calcite: $CaCO_3$ Hexagonal

Rhodonite is a mineral, like rhodochrosite, that commonly displays the pink of certain manganese compounds. The name comes from the Greek *rhodon* (rose).

148. ROSASITE and LIMONITE

Collection: Court
Size: $6\frac{1}{2}\times6$ in. (17×15 cm.)
Locality: Ojuela Mine, Mapimí, Durango, Mexico
Rosasite: $(Cu,Zn)_2CO_3(OH)_2$ Monoclinic
Limonite (goethite): $FeO(OH)$ Orthorhombic

The blue-green color of rosasite is similar to that of many secondary copper minerals. The name comes from the Rosas Mine in Sardinia, where the mineral was first discovered.

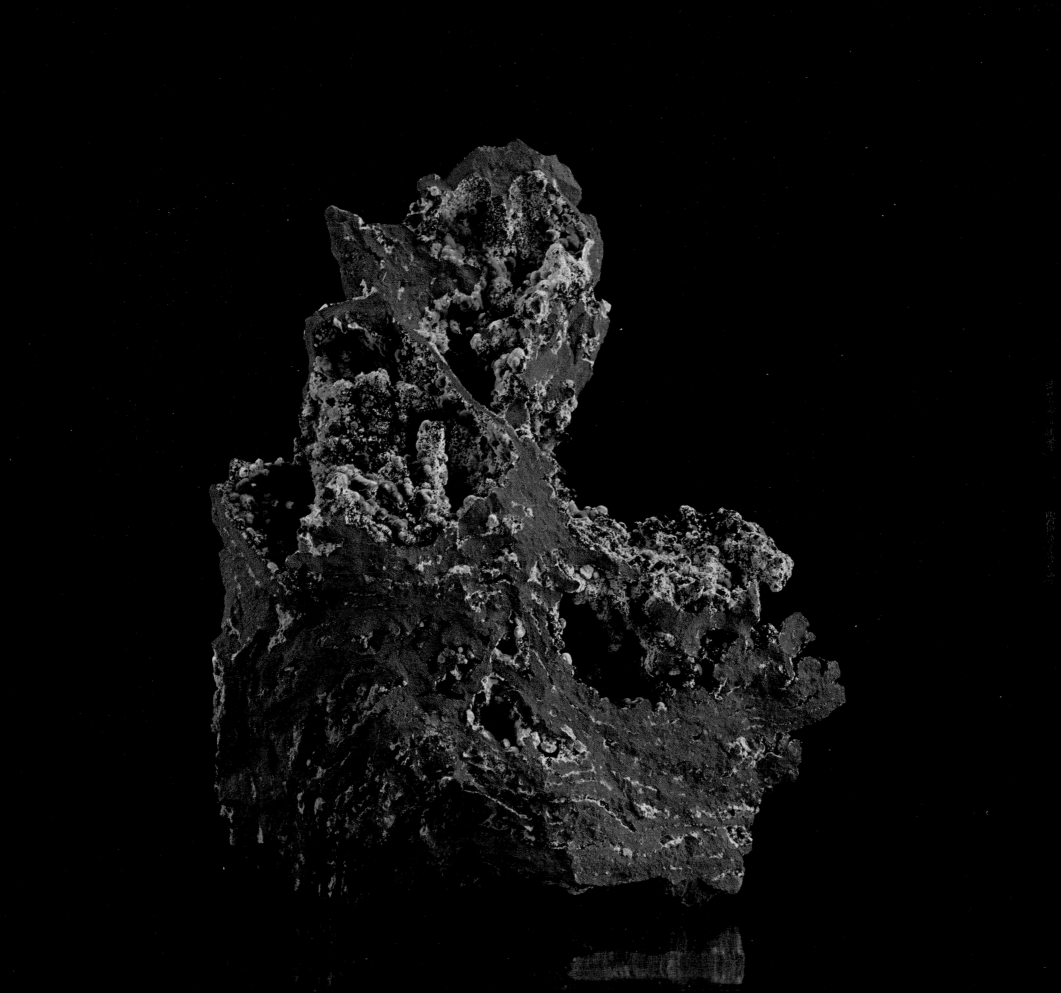

149. RUTILE in QUARTZ

Collection: Court
Size: $4\frac{1}{2} \times 4\frac{1}{2}$ in. (11×11 cm.)
Locality: Minas Gerais, Brazil
Rutile: TiO_2 Tetragonal
Quartz: SiO_2 Hexagonal

These hairlike crystals of rutile, small as they are, are still too large to produce the light-scattering effect responsible for rose quartz (see plate 142). Specimens such as this, where the rutile is visible, are known as rutilated quartz. Rutile sometimes accumulates in significant amounts in black sands on beaches (that is, beach placers) and becomes an important ore of titanium. Rutile gets its name from the Latin *rutilus* (red, or reddish) because of the reddish color of the more translucent varieties.

150. SCOLECITE

Collection: Halpern
Size: $11\frac{1}{2} \times 6$ in. (29 × 15 cm.)
Locality: Soledade, Rio Grande do Sul, Brazil
$Ca(Al_2Si_3)O_{10} \cdot 3H_2O$ Monoclinic

Scolecite is another member of the zeolite group (see, for example, plate 97). When heated, it fuses easily and has a tendency to curl up like a worm—hence its name, which comes from the Greek *skolhex* (worm).

151. SEMSEYITE and QUARTZ

Collection: McGuinness
Size: $3 \times 2\frac{1}{2}$ in. (8×6 cm.)
Locality: Herja, Rumania
Semseyite: $Pb_9Sb_8S_{21}$ Monoclinic
Quartz: SiO_2 Hexagonal

Semseyite was named for Andor von Semsey (died 1923), a Hungarian
nobleman who was interested in minerals.

152. SIDERITE on drusy ALBITE

Collection: Court
Size: $8 \times 8\frac{1}{2}$ in. (20×22 cm.)
Locality: Saint Hilaire, Quebec, Canada
Siderite: $FeCO_3$ Hexagonal
Albite: $NaAlSi_3O_8$ Triclinic

The rhombohedral crystal form, as well as the rhombohedral cleavage,
of siderite are well shown by this specimen. The mineral has at times
been a source of iron, since the metal is easily recoverable from it. The
name comes from the Greek *sideros* (iron).

153. SIDERITE and QUARTZ

Collection: Court
Size: 11 × 11 in. (28 × 28 cm.)
Locality: Moro Velho Mine, Minas Gerais, Brazil
Siderite: $FeCO_3$ Hexagonal
Quartz: SiO_2 Hexagonal

This unusual specimen provides an interesting contrast between the very different habits (see the Introduction) of two hexagonal minerals. Siderite is in rhombs of very shallow slope, which give a somewhat scaly appearance to the crystal aggregate. Quartz exhibits its typical hexagonal prisms, capped by two steep rhombohedrons which are so evenly developed that together they appear to be a single hexagonal pyramid.

154. SILVER

Collection: Halpern
Size: $2\frac{1}{2} \times 3\frac{1}{2}$ in. (6×9 cm.)
Locality: Kongsberg, Norway
Ag   Isometric

Silver, although a more abundant metal in the earth's crust than gold, occurs much more rarely as the native element. Specimens such as this are sometimes called wire silver. The "wires" are cubic crystals distorted and greatly elongated along one axis.

The origin of the name is all but lost in antiquity. It may have been derived from the Akkadian (Babylonian) *sarapu* (to smelt).

155. SKLODOWSKITE on GYPSUM

Collection: Court
Size: 3 × 4 in. (8 × 10 cm.)
Locality: Chihuahua, Mexico
Sklodowskite: $Mg(UO_2)_2Si_2O_7 \cdot 6H_2O$ Orthorhombic
Gypsum: $CaSO_4 \cdot 2H_2O$ Monoclinic

Pale-yellow (a color characteristic of many uranium minerals) crystals of sklodowskite are perched on white gypsum. The mineral name honors Maria Sklodowska Curie (1867–1934), Nobel Prize winner and discoverer of radium.

156. SMITHSONITE

Collection: Land
Size: $5\frac{1}{2}$ × 7 in. (14 × 18 cm.)
Locality: Tsumeb, South-West Africa
$ZnCO_3$ Hexagonal

Smithsonite is most often grayish white to colorless, although pale-green, blue, and yellow colors are not uncommon. The pink color displayed here is very rare. It may be due to traces of cobalt. The name honors the English chemist and mineralogist James Smithson (1765–1829), who left a substantial bequest for founding the Smithsonian Institution in Washington, D.C.

157. SPHALERITE with ARSENOPYRITE

Collection: Krueger
Size: $1\frac{1}{2} \times 5\frac{1}{2}$ in. $(4 \times 14$ cm.$)$
Locality: Trepca, Yugoslavia
Sphalerite: $(Zn,Fe)S$ Isometric
Arsenopyrite: $FeAsS$ Orthorhombic

The sphalerite crystals are dominated by the mineral's characteristic tetrahedral form. The name is from the Greek *sphaleros* (slippery, or deceptive), because the mineral often misled early miners into thinking they had an ore of lead, or of some other metal, whereas they ended up getting nothing because zinc is not easy to recover, as are lead, copper, or silver.

158. STAUROLITE

Collection: Court
Size: $3 \times 4\frac{1}{2}$ in. $(8 \times 11$ cm.$)$
Locality: Minas Gerais, Brazil
$(Fe,Mg)_2Al_9Si_4O_{23}(OH)$ Monoclinic

A penetration twin of staurolite is shown here embedded in mica schist, a metamorphic rock in which the mineral is frequently found. The mineral may form twins, sometimes interpenetrating at right angles, sometimes (as here) in an oblique pattern; in either case, such natural crosses were regarded with wonder and desire, and were highly prized by early Christians, who considered them charms with which to ward off the devil. Such staurolite crosses still enjoy a considerable trade as amulets and good-luck stones. The name comes from the Greek *stauros* (a cross).

159. STIBICONITE

Collection: Court
Size: 7 × 7 in. (18 × 18 cm.)
Locality: La Bufa Charcas Mine, San Luis Potosí State, Mexico
$Sb_3O_6(OH)$ Isometric

The radiating crystals are not characteristic of an isometric mineral, thus suggesting that this interesting specimen is actually a pseudomorph (see the Introduction) of stibiconite after some other mineral. The name comes from the Latin *stibium* (antimony) and the Greek *koris* (powder), because the mineral is commonly found in powdery form.

160. STIBNITE

Collection: Land
Size: $11 \times 8\frac{1}{2}$ in. (28×22 cm.)
Locality: Baia-Sprie, Rumania
Sb_2S_3 Orthorhombic

Long before this mineral was recognized as an ore of antimony, it was used by ancient Greek and Phoenician women for darkening their eyelashes and for other cosmetic applications, to which the mineral's intense blackness and softness ($H=2$) were well suited. Named for the Latin *stibium* (antimony).

161. STIBNITE

Collection: Land
Size: $11 \times 8\frac{1}{2}$ in. (28×22 cm.)
Locality: Ichikonawa Mine, Shikoku, Japan
Sb_2S_3 Orthorhombic

Crystals of stibnite with stature such as this one are relatively rare. Japan is one of the few countries that has been a prime source in the past for such quality specimens.

162. STILBITE and APOPHYLLITE

Collection: Halpern
Size: $5 \times 2\frac{1}{2}$ in. (13×16 cm.)
Locality: Poona district, India
Stilbite: $NaCa_2(Al_5Si_{13})O_{36} \cdot 14H_2O$ Monoclinic
Apophyllite: $KCa_4Si_8O_{20}(F,OH) \cdot 8H_2O$ Tetragonal

Radiating crystals of stilbite, a mineral of the zeolite group (see, for example plate 97), are associated here with apophyllite (see also plates 10, 11, 162), which is often hard to see (whether in a photograph or in real life) because of its very low luster (low index of refraction). Apophyllite, although not a true zeolite, is often associated with, and has many characteristics in common with, the zeolite group.

The name stilbite stems from the Greek *stilbos* (shimmering), in reference to the appearance of the crystal aggregates.

163. SULFUR (with drusy coating of CELESTITE)

Collection: Court
Size: $6\frac{1}{2} \times 4$ in. (17×10 cm.)
Locality: Province of Agrigento, Sicily
Sulfur: S   Orthorhombic
Celestite: $SrSO_4$ Orthorhombic

Sulfur is one of the few minerals that is not a chemical compound but an element, found as such in the crust of the earth like gold, silver, and copper. But unlike these other elements, which are metals, sulfur is—along with carbon—a native nonmetal.

Somewhat surprisingly, in view of the very limited occurrences of most native elements in the crust of the earth, sulfur is occasionally found in deposits of millions of tons. Today, these constitute one of the principal sources for industrial sulfur and all of its compounds. This is fortunate for man, since sulfur is one of the most important and most versatile materials in the mineral world. In fact, some economists and sociologists have made a good case for using the sulfur index (that is, the per-capita consumption of sulfur) as the best single measure of the stage of industrial development of a nation; in other words, those nations which use the most sulfur, per capita, presumably represent those enjoying the most advanced industrial development.

Sulfur is, for example, one of the principal sources of sulfuric acid (H_2SO_4)—a staple of the entire chemical industry, where its applications are manifold. Refined elemental sulfur is used in pesticides, medicines, soil amendments, and many other applications.

Sulfur is one of the very few minerals that can readily be ignited and can burn completely. In times past it has even been used as a fuel. It was thus employed to melt and extract deposits of the element on the island of Sicily, which was for many years the chief supplier of sulfur to the world. Wasteful and noxious as this practice may now seem (the burning of sulfur generates quantities of poisonous sulfur-dioxide gas), it was in its time practical, economic, and acceptable.

Today sulfur is mined chiefly by the Frasch process, named after the chemical engineer who, around the turn of the century, developed a successful method for extracting sulfur known to be associated—at depths of several hundred feet–with some of the salt domes along the Gulf Coast of the United States. For many years previous to the Frasch process, attempts to produce the sulfur by conventional shaft-sinking and mining methods had proved unsuccessful, and even disastrous to life. In the Frasch process, a well is sunk to the sulfur-bearing stratum; superheated steam which melts the sulfur is forced down; then, with an airlift, it is brought to the surface where, still molten, it is piped into tank cars and even into insulated and heated tanker-ships which transport it to European ports, where special facilities for unloading and distribution have been built. Sulfur thus shares the distinction, along with mercury, of being regularly transported long distances in liquid form. And sulfur is unusual in yet other ways. For example, the highest mining operation ever undertaken was at the Aucanquilcha Mine, near the summit of the volcano with the same name, in Chile, at an elevation of just over 20,000 feet. This was a sulfur mine which operated successfully for several years, despite the difficult working conditions at that extreme altitude. Eventually, in part because of exhaustion of the higher-grade ores, it closed down in 1967.

The name of the mineral comes directly from the Latin *sulfur*.

164. SULFUR on GYPSUM (variety selenite)

Collection: Court
Size: $5 \times 7\frac{1}{2}$ in. (13×19 cm.)
Locality: Province of Agrigento, Sicily
Sulfur: S Orthorhombic
Gypsum: $CaSO_42H_2O$ Monoclinic

171. TOURMALINE with QUARTZ

Collection: Court
Size: 12×11 in. $(30 \times 28$ cm.$)$
Locality: Minas Gerais, Brazil
Tourmaline: $(Na,Ca)(Mg,Fe^{+2},Fe^{+3},Al,Li)_3Al_6(BO_3)_3Si_6$
               $O_{18}(OH)_4$ Hexagonal
Quartz: SiO_2 Hexagonal

A good deal of mineralogical history is displayed in this specimen. A large single crystal of quartz grew in its typical hexagonal prismatic habit. At some point, the character of the surrounding solutions changed and locally dissolved away much of the interior of the quartz crystal, while leaving some of the external prism faces untouched. Then began deposition of material of much greater chemical complexity— that is, the green, gem-quality tourmaline whose formula approximates $(Na,Mg)(Al,Fe,Li,Mg)_3B_3Al_3(AlSi_2O_9)_3(O,OH,F)_4$. Tourmaline, in fact, possesses perhaps the most varied and complex chemistry of any mineral. That the crystal structure (hexagonal) of tourmaline can accommodate such a variety of atoms, often in varying proportions, is of considerable interest, as is the presence in the formula of such volatile components as fluorine, boron, and hydroxyl, as well as the rather rare, light metal, lithium. The green color is at least in part the result of changes in the ratio of ferrous (Fe^{++}) to ferric (Fe^{+++}) iron content.

172. VANADINITE

Collection: Land
Size: $6\frac{1}{2} \times 8$ in. (17×20 cm.)
Locality: Old Yuma Mine, Tucson, Arizona
$Pb_5(VO_4)_3Cl$ Hexagonal

Compare these beautifully formed hexagonal crystals with those of mimetite (plates 109, 110) and pyromorphite (plate 125). Except for the fact that mimetite is white to yellow, pyromorphite is green, and vanadinite is red, they are all nearly identical. Yet mimetite is an arsenate, pyromorphite is a phosphate, and vanadinite is a vanadate. These three minerals constitute one of the more remarkable cases where substitution of the acid radical (rather than the more common case of substitution of metal cations, as magnesium for iron, or sodium for potassium) can take place without distortion of crystal structure. The name vanadinite is in recognition of the vanadium content of the mineral. The name of the element vanadium is from Old Norse, Vanada, the Scandinavian goddess of fertility, now known to us under the name Freya.

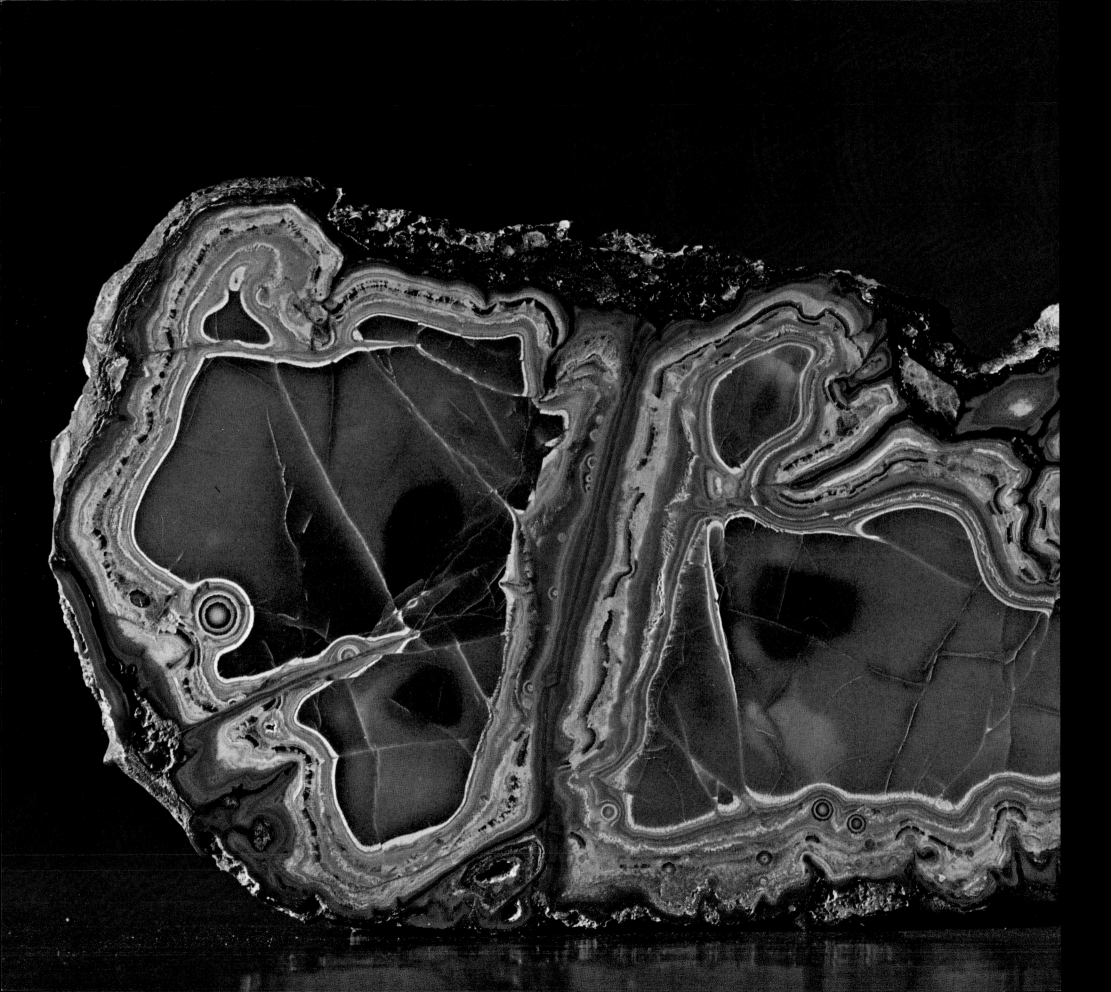

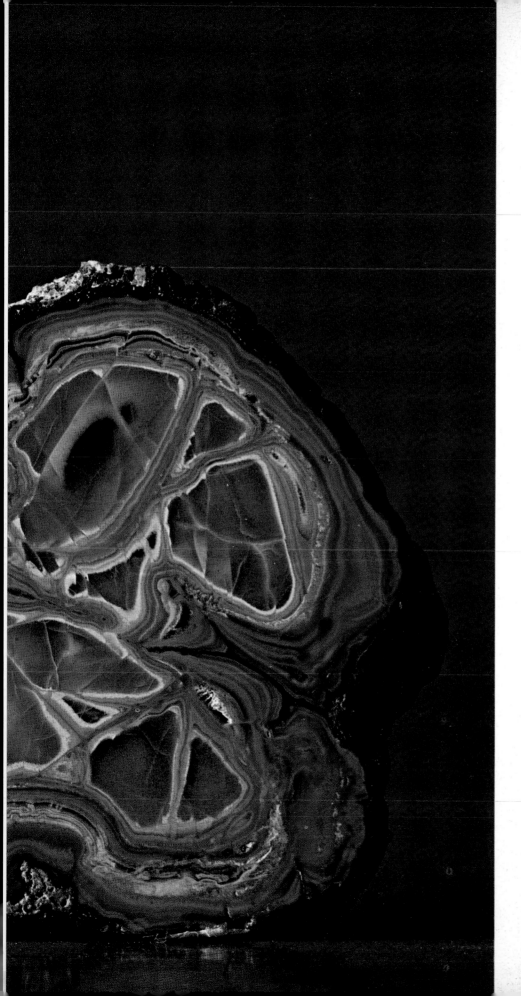

173. VARISCITE with CRANDALLITE and
　　　WARDITE

Collection:　McGuinness
Size:　　　　$6 \times 2\frac{1}{2}$ in. (15×6 cm.)
Locality:　　Fairfield, Utah County, Utah
Variscite: $AlPO_4 \cdot 2H_2O$　Orthorhombic
Crandallite: $CaAl_3(PO_4)_2(OH)_5 \cdot H_2O$　Hexagonal
Wardite: $NaAl_3(PO_4)_2(OH)_4 \cdot 2H_2O$　Tetragonal

The dark-green central portions of this polished slab are massive variscite; these are rimmed by paler-green wardite, and in turn by yellow to white crandallite.

　Variscite takes its name from Variscia, an old name for a district in Saxony where the mineral was first discovered. Crandallite was named for M. L. Crandall, and wardite for Henry A. Ward, founder of Ward's Scientific Establishment in Rochester, N.Y., and an early and well-known mineral dealer and supplier of natural-history specimens to the scientific establishment.

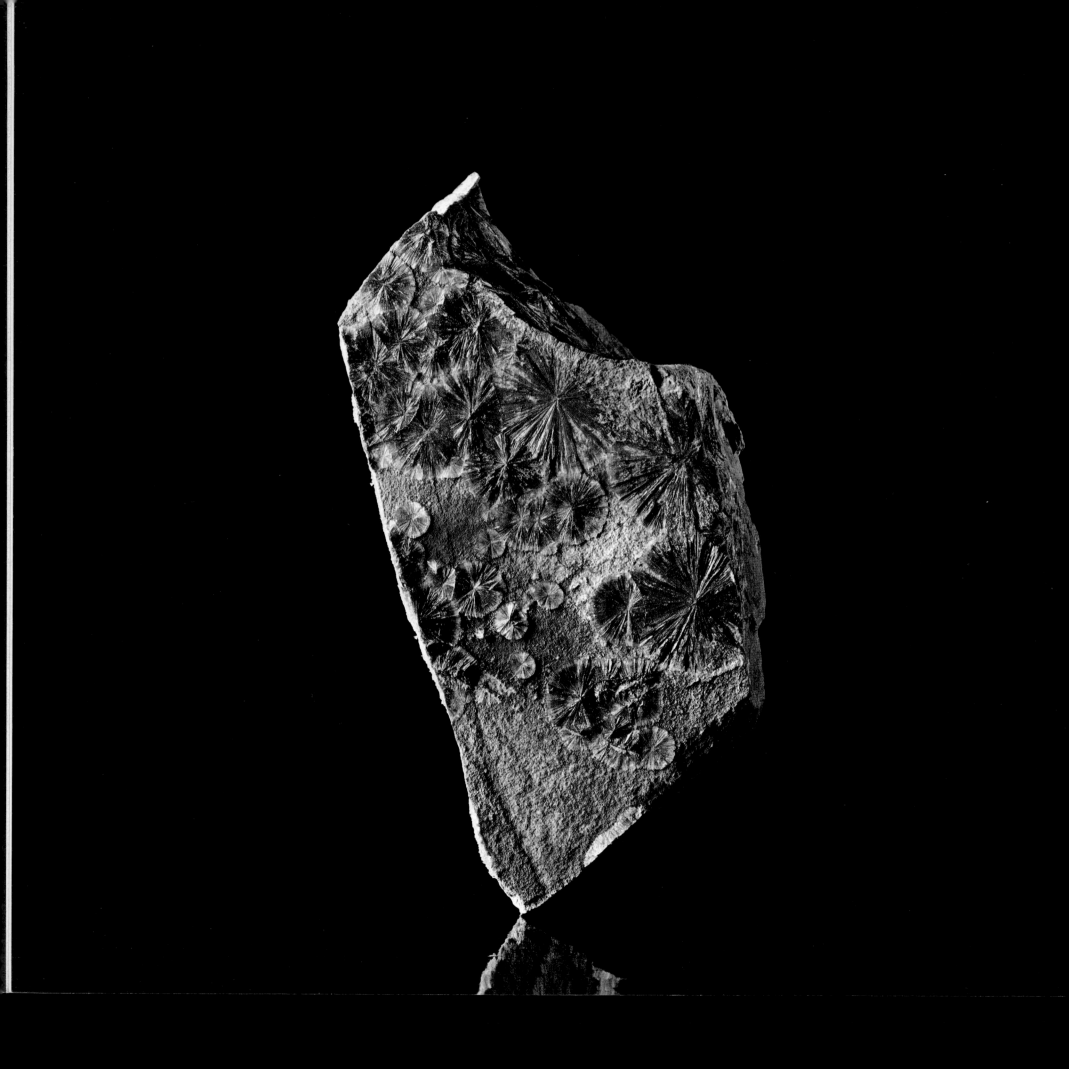

177. WAVELLITE

Collection: Land
Size: 3×4 in. $(8 \times 10$ cm.)
Locality: Hot Springs, Arkansas
$Al_3(PO_4)_2(OH)_3 \cdot 5H_2O$ Orthorhombic

The mineral here displays well its characteristic stellate growth pattern. Deposits in Pennsylvania were at one time exploited as a source of phosphorus. The name honors William Wavell (died 1829), an English physician, who first discovered the mineral.

178. WILLEMITE

Collection: McGuinness
Size: $1\frac{1}{2} \times 2\frac{1}{2}$ in. (4 × 6 cm.)
Locality: Berg Aukas Mine, Grootfontein, South-West
              Africa

Zn_2SiO_4 Hexagonal

The mineral was named by the French mineralogist Lévy, in honor of
William I, King of the Netherlands.

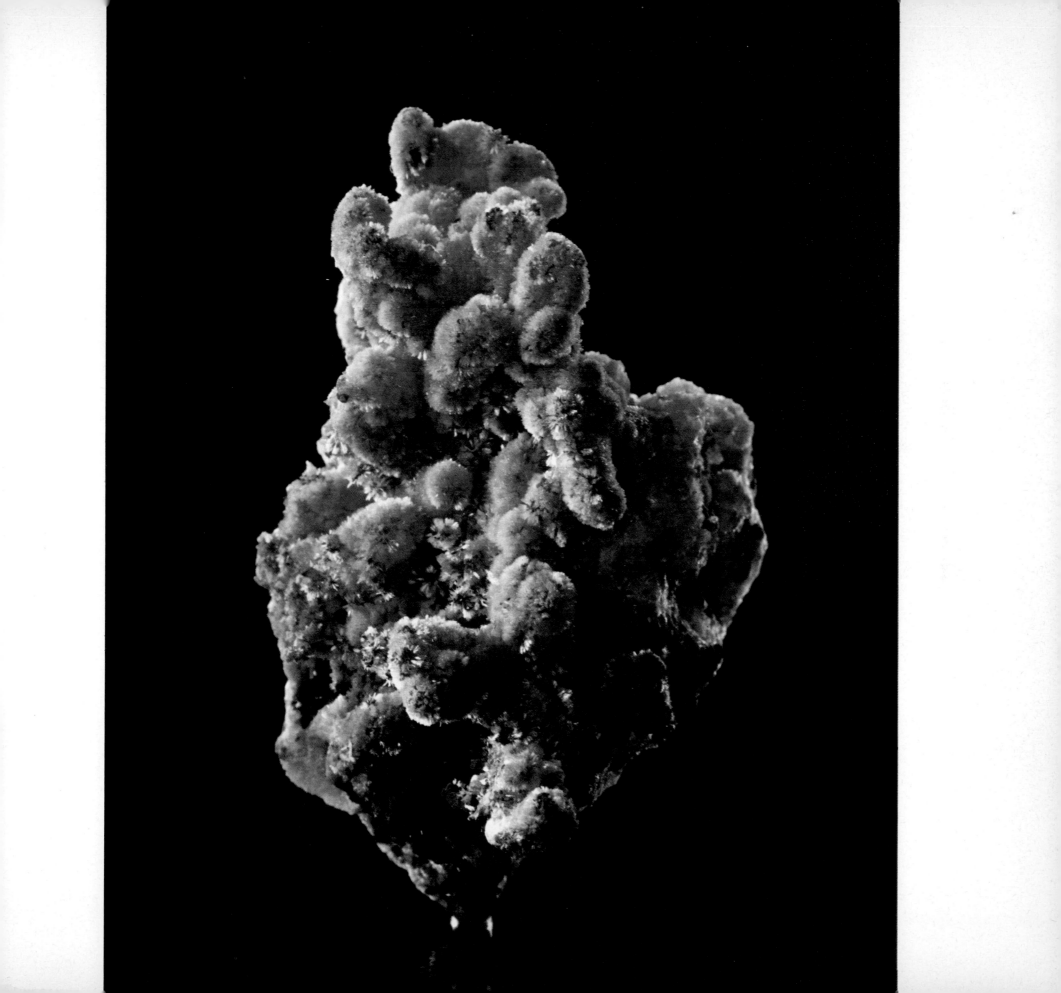

179. WILLEMITE with FRANKLINITE and
CALCITE (photographed without filtering
out of ultraviolet)

Collection: Krueger
Size: $2\frac{1}{2} \times 2\frac{1}{2}$ in. (6×6 cm.)
Locality: Franklin Furnace, New Jersey
Willemite: Zn_2SiO_4 Hexagonal
Franklinite: $(Zn,Mn^{+2},Fe^{+2})(Fe^{+3},Mn^{+3})_2O_4$ Isometric
Calcite: $CaCO_3$ Hexagonal

The unusual colors displayed here are the result of exposing the specimen to fluorescent light, to which the ordinarily white willemite (see plate 178) responds with a brilliant color display, supplemented by the calcite. The franklinite is in black, metallic grains and is known to occur in only one locality in the world (Franklin Furnace, New Jersey—from which the name franklinite is derived). In fact, it is upon this deposit that the New Jersey Zinc Company founded its fortunes.

180. WOLFRAMITE and QUARTZ

Collection: Court
Size: $4\frac{1}{2} \times 4\frac{1}{4}$ in. (11×11 cm.)
Locality: Panasqueira, Portugal
Wolframite: $(Fe,Mn)WO_4$ Monoclinic
Quartz: SiO_2 Hexagonal

Wolframite (see also ferberite and huebnerite) is one of the principal ores of tungsten. The origin of the name (*Wolfram* in German) is lost in antiquity. It has been suggested that, since the root *wolf* refers to a feared animal of medieval and earlier times, the name may have been applied to the mineral that "devoured" (that is, decreased) the yield of tin, in the tin ores in which it is frequently present. In other words, wolframite was an objectional component, for tungsten was unknown to the early miners and wolframite interfered with concentration and smelting of cassiterite in the obtaining of tin.

181. WULFENITE

Collection: Court
Size: $7 \times 7\frac{1}{2}$ in. (18×19 cm.)
Locality: Sierra de los Lamentos, Chihuahua, Mexico
$PbMoO_4$ Tetragonal

The tetragonal form, involving a small four-sided prism and a large two-faced basal pinacoid, is here beautifully displayed. The name honors Franz X. von Wulfen (1728–1805), an Austrian mineralogist.

182. WULFENITE

Collection: Smale
Size: $4\frac{1}{2} \times 5\frac{1}{2}$ in. (11×14 cm.)
Locality: Globe-Miami mining district, Gila County, Arizona
$PbMoO_4$ Tetragonal

These crystals are so thin that the prism face almost disappears and the basal pinacoid becomes the dominant form.

183. AN ASSORTMENT OF CHOICE SPECIMENS
Collection: Court

From left to right:
SULFUR
   Locality: San Felipe, Baja California, Mexico
GYPSUM, variety selenite (foreground)
   Locality: Jackson County, Texas
   The color results from a thin coating of brown limonite.
FLUORITE and CALCITE (background)
   Locality: Cave in Rock, Illinois
   In this specimen tiny scalenohedrons of white calcite are
   perched on a cube of purple fluorite.
AZURITE on MALACHITE (background)
   Locality: Bisbee, Arizona
PYRITE (foreground)
   Locality: Niaca, Chihuahua, Mexico
   This specimen shows the pyritohedral crystal form.
Brassy yellow CHALCOPYRITE and black SPHALER-
ITE crystals, with white QUARTZ (background)
   Locality: Niaca, Chihuahua, Mexico
QUARTZ, variety amethyst, on COPPER (foreground)
   Locality: Vera Cruz State, Mexico
GYPSUM, variety selenite (background)
   Locality: Cave of Swords, Chihuahua, Mexico
Two halves of a QUARTZ geode (foreground)
   The first layer inside the rough exterior is bluish white
   chalcedony (agate); the next layer is quartz (rock crys-
   tal) whose terminations (innermost layer) grade into
   amethyst.
TOURMALINE, gem-quality crystal (background)
   Locality: Santa Rosa Mine, Minas Gerais, Brazil
QUARTZ sandstone of the Green River formation (back-
ground)
   Locality: Farson, Wyoming
   The quartz grains are stained with limonite, which gives
   the sandstone its yellow to brown color. The sand grains
   enclose a fossil fish. The Green River formation belongs
   to the Eocene epoch, some forty to fifty million years ago.
QUARTZ, rock crystal with traces of amethyst
   Locality: Vera Cruz State, Mexico

Appendix:
Symbols of the Chemical Elements

The seeming lack of correspondence between certain symbols and their names (such as Au for gold, K for potassium, Na for sodium) results from the fact that the symbols are in many cases derived from the Latin rather than the English names for the elements. Thus Au comes from *aurum* (gold); K, from *kalium* (potassium); and Na, from *natron* (sodium).

For convenience in looking up a symbol, the tabulation is alphabetic by symbol, not by the name of the element.

Symbol	Chemical Element	Symbol	Chemical Element	Symbol	Chemical Element
Ac	Actinium	Dy	Dysprosium	Mo	Molydenum
Ag	Silver	E	Einsteinium	Mv	Mendelevium
Al	Aluminum	Er	Erbium	N	Nitrogen
Am	Americium	Eu	Europium	Na	Sodium
Ar	Argon	F	Fluorine	Nb	Niobium
As	Arsenic	Fe	Iron	Nd	Neodymium
At	Astatine	Fm	Fermium	Ne	Neon
Au	Gold	Fr	Francium	Ni	Nickel
B	Boron	Ga	Gallium	Np	Neptunium
Ba	Barium	Gd	Gadolinium	O	Oxygen
Be	Beryllium	Ge	Germanium	Os	Osmium
Bi	Bismuth	H	Hydrogen	P	Phosphorus
Bk	Berkellium	He	Helium	Pa	Protactinium
Br	Bromine	Hf	Hafnium	Pb	Lead
C	Carbon	Hg	Mercury	Pd	Palladium
Ca	Calcium	Ho	Holmium	Pm	Promethium
Cd	Cadmium	I	Iodine	Po	Pollonium
Ce	Cerium	Ir	Iridium	Pt	Platinum
Cf	Californium	K	Potassium	Pu	Plutonium
Cl	Chlorine	Kr	Krypton	Ra	Radium
Cm	Curium	La	Lanthanum	Rb	Rubidium
Co	Cobalt	Li	Lithium	Re	Rhenium
Cr	Chromium	Lu	Lutetium	Rh	Rhodium
Cs	Cesium	Mg	Magnesium	Rn	Radon
Cu	Copper	Mn	Manganese	Ru	Ruthenium

Symbol	Chemical Element	Symbol	Chemical Element	Symbol	Chemical Element
S	Sulfur	Sr	Strontium	Tl	Thallium
Sb	Antimony	Ta	Tantalum	Tm	Thulium
Sc	Scandium	Tb	Terbium	U	Uranium
Se	Selenium	Tc	Technetium	V	Vanadium
Si	Silicon	Te	Tellurium	W	Tungsten
Sm	Samarium	Th	Thorium	Xe	Xenon
Sn	Tin	Ti	Titanium		

Selected Bibliography

This list of recommended reading is in no sense exhaustive, but it includes enough variety and specialization to satisfy the needs of most persons interested in knowing more about minerals and rocks.

Bancroft, Peter, *The World's Finest Minerals and Crystals*, 1973

Boegel, Hellmuth (John Sinkakas, translator), *The Studio Handbook of Minerals*, 1972

Dake, H. C. *et al*, *Quartz Family Minerals*, 1938

Desautels, Paul E., *The Mineral Kingdom*, 1968

English, George Letchworth, *Getting Acquainted with Minerals*, 1934

Fenton, Carrol Lane, and Fenton, Mildred Adams, *The Rock Book*, 1940

Hurlbut, Cornelius S., Jr., *Minerals and Man*, 1969

Kunz, George Frederick, *Gems and Precious Stones of North America*, 1892; Dover edition, 1968

Loomis, Frederic B., *Field Book of Common Rocks and Minerals*, revised edition, 1948

Metz, Rudolph, *Precious Stones and Other Crystals*, 1965

Nicolay, H. V. and Stone, A. V., *Rocks and Minerals: a Guide for Collectors of the Eastern United States*, 1967

Pearl, Richard M., *American Gem Trails*, 1964

——, *Rocks and Minerals*, 1956

Pough, Frederick, *A Field Guide to Rocks and Minerals*, third edition, 1970

Rapp, George, Jr., *Color of Minerals*, 1971

Sinkakas, John, *Gemstones and Minerals: How and Where to Find Them*, 1961

——, John, *Mineralogy for Amateurs*, 1964

Sosman, R. B., *The Properties of Silica*, 1927

Vanders, Iris, and Kerr, Paul F., *Mineral Recognition*, 1967

Whitlock, Herbert P., *The Story of the Gems*, 1946

Wilson, Mab, *Gems*, 1967

Some Standard References

A few works on mineralogy are so complete, so thorough, and so widely used that they should not be omitted from any reference list on minerals, whether for amateurs or professionals. The ones cited below have in fact been used extensively in obtaining and in checking much of the data presented in this book; our debt to these authors is large.

The System of Mineralogy of James D. Dana and Edward S. Dana, seventh edition
vol. I: Charles Palache, Harry Berman,

and Clifford Frondel, *Elements, Sulfides, Sulfosalts, Oxides*, 1944
vol. II: Charles Palache, Harry Berman,

and Clifford Frondel, *Halides, Nitrates, Borates, Carbonates, Sulfates, Phosphates, Arsenates, Tungstates, Molybdates, Etc.,* 1951
vol. III: Clifford Frondel, *Silica Minerals,* 1962

Hey, M. H., *Chemical Index of Minerals,* second edition, 1962
Strunz, Hugo, *Mineralogische Tabellen,* fifth edition, 1970

Periodicals

In addition to such books as those listed above, a number of periodicals are devoted to the interests of mineral collectors and students, and will provide rewarding reading on a wide diversity of topics as well as reporting matters of current interest. Some of the journals that can be recommended are:

Earth Science
Gems and Minerals
Lapidary Journal
Mineralogical Record
Rocks and Minerals

Index of Illustrated Minerals